Analysis of Methods to Determine Storage Capacity of, and Sedimentation in, Loch Lomond Reservoir, Santa Cruz County, California, 2009

By Kelly R. McPherson, Lawrence A. Freeman, and Lorraine E. Flint

Prepared in cooperation with the City of Santa Cruz

Scientific Investigations Report 2011–5141

U.S. Department of the Interior
U.S. Geological Survey

U.S. Department of the Interior
KEN SALAZAR, Secretary

U.S. Geological Survey
Marcia K. McNutt, Director

U.S. Geological Survey, Reston, Virginia: 2011

For more information on the USGS—the Federal source for science about the Earth, its natural and living resources, natural hazards, and the environment, visit http://www.usgs.gov or call 1–888–ASK–USGS.

For an overview of USGS information products, including maps, imagery, and publications, visit http://www.usgs.gov/pubprod

To order this and other USGS information products, visit http://store.usgs.gov

Suggested citation:
McPherson, K.R., Freeman, L.A., and Flint, L.E., 2011, Analysis of Methods to Determine Storage Capacity of, and Sedimentation in, Loch Lomond Reservoir, Santa Cruz County, California, 2009: U.S. Geological Survey Scientific Investigations Report 2011–5141, 88 p.

Contents

Figures

Tables

Conversion Factors

Inch/Pound to SI

Multiply	By	To obtain
Length		
foot (ft)	0.3048	meter (m)
yard (yd)	0.9144	meter (m)
mile (mi)	1.609	kilometer (km)
Area		
acre	4,047	square meter (m²)
acre	0.004047	square kilometer (km²)
square foot (ft²)	0.09290	square meter (m²)
square mile (mi²)	2.590	square kilometer (km²)
Volume		
cubic inches (in³)	16.39	cubic centimeters (cm³)
cubic foot (ft³)	0.02832	cubic meter (m³)
acre-foot (acre-ft)	1,233	cubic meter (m³)
Flow rate		
acre-foot per day (acre-ft/d)	0.01427	cubic meter per second (m³/s)
acre-foot per year (acre-ft/yr)	1,233	cubic meter per year (m³/yr)
cubic foot per second (ft³/s)	0.02832	cubic meter per second (m³/s)
cubic foot per day (ft³/d)	0.02832	cubic meter per day (m³/d)
Mass		
pound (lb)	0.4536	kilogram
ton per day (ton/d)	0.9072	metric ton per day
ton per year (ton/yr)	0.9072	metric ton per year
ton per year per mile squared	0.3503	metric ton per year per kilometer squared

SI to Inch/Pound

Multiply	By	To obtain
Length		
decimeter (dm)	3.9370	inch (in.)
meter (m)	3.2808	feet (ft)
kilometer (km)	0.6214	mile (mi)

Notes:

Vertical coordinate information is referenced to the National Geodetic Vertical Datum of 1929 (NGVD 29).

Horizontal coordinate information is referenced to the North American Datum of 1983 (NAD 83).

Altitude, as used in this report, refers to distance above the vertical datum.

Analysis of Methods to Determine Storage Capacity of, and Sedimentation in, Loch Lomond Reservoir, Santa Cruz County, California, 2009

By Kelly R. McPherson, Lawrence A. Freeman, and Lorraine E. Flint

Abstract

In 2009, the U.S. Geological Survey, in cooperation with the City of Santa Cruz, conducted bathymetric and topographic surveys to determine the water storage capacity of, and the loss of capacity owing to sedimentation in, Loch Lomond Reservoir in Santa Cruz County, California. The topographic survey was done as a supplement to the bathymetric survey to obtain information about temporal changes in the upper reach of the reservoir where the water is shallow or the reservoir may be dry, as well as to obtain information about shoreline changes throughout the reservoir. Results of a combined bathymetric and topographic survey using a new, state-of-the-art method with advanced instrument technology indicate that the maximum storage capacity of the reservoir at the spillway altitude of 577.5 feet (National Geodetic Vertical Datum of 1929) was 8,646 ±85 acre-feet in March 2009, with a confidence level of 99 percent. This new method is a combination of bathymetric scanning using multibeam-sidescan sonar, and topographic surveying using laser scanning (LiDAR), which produced a 1.64-foot-resolution grid with altitudes to 0.3-foot resolution and an estimate of total water storage capacity at a 99-percent confidence level. Because the volume of sedimentation in a reservoir is considered equal to the decrease in water-storage capacity, sedimentation in Loch Lomond Reservoir was determined by estimating the change in storage capacity by comparing the reservoir bed surface defined in the March 2009 survey with a revision of the reservoir bed surface determined in a previous investigation in November 1998. This revised reservoir-bed surface was defined by combining altitude data from the 1998 survey with new data collected during the current (2009) investigation to fill gaps in the 1998 data. Limitations that determine the accuracy of estimates of changes in the volume of sedimentation from that estimated in each of the four previous investigations (1960, 1971, 1982, and 1998) are a result of the limitations of the survey equipment and data-processing methods used. Previously used and new methods were compared to determine the recent (1998–2009) change in storage capacity and the most accurate and cost-effective means to define the reservoir bed surface so that results can be easily replicated in future surveys.

Results of this investigation indicate that the advanced method used in the 2009 survey accurately captures the features of the wetted reservoir surface as well as features along the shoreline that affect the storage capacity calculations. Because the bathymetric and topographic data are referenced to a datum, the results can be easily replicated or compared with future results. Comparison of the 2009 reservoir-bed surface with the surface defined in 1998 indicates that sedimentation is occurring throughout the reservoir. About 320 acre-feet of sedimentation has occurred since 1998, as determined by comparing the revised 1998 reservoir-bed surface, with an associated maximum reservoir storage capacity of 8,965 acre-feet, to the 2009 reservoir bed surface, with an associated maximum capacity of 8,646 acre-feet. This sedimentation is more than 3 percent of the total storage capacity that was calculated on the basis of the results of the 1998 bathymetric investigation.

Introduction

Loch Lomond Reservoir, an impoundment of Newell Creek, is in the Santa Cruz Mountains, California (fig. 1), and is owned by the City of Santa Cruz. Newell Creek, a tributary of the San Lorenzo River, was dammed about 9 mi north of the City of Santa Cruz to create the reservoir as a source of water supply for the residents of the City of Santa Cruz. Runoff from a watershed of more than 8 mi^2 is now stored and is available for public use. The Newell Creek Dam is an earth-fill barricade, 190 ft high and 750 ft long, that was completed in 1960 (fig. 2). Impounded water first flowed over the spillway in March 1963.

Figure 1.   Beginning stages of the construction of the Newell Creek Dam in the Santa Cruz Mountains, looking upstream of the Loch Lomond Recreation Area, Santa Cruz County, California,1960. Photograph from the City of Santa Cruz, Graham Hill Water Treatment Plant.

Since the reservoir was completed, park rangers and city water managers have routinely observed sedimentation at the inflow of Newell Creek. Water managers for the City of Santa Cruz periodically measure storage capacity to determine whether sedimentation has occurred, allowing them to take timely and appropriate action to regulate water supply. Sedimentation occurs in the lower reach of the reservoir because of landslides from the surrounding slopes, and in the upstream reach because of the inflow of sediment carried in streamflow from Newell Creek (Fogelman and Johnson, 1985).

In 2009, the U.S. Geological Survey (USGS) began an investigation in cooperation with the City of Santa Cruz and with the assistance of California State University Monterey Bay (CSUMB) Seafloor Mapping Lab (SFML) to determine the current storage capacity of Loch Lomond Reservoir, and to compare the results of this investigation with those of the four previous similar investigations in order to estimate the loss of capacity owing to sedimentation. For the purposes of this comparison, the volume of sedimentation in the reservoir is considered equal to the decrease in water storage capacity; storage capacity is the volume above the reservoir-bed surface to the altitude of the spillway. To determine recent sedimentation in Loch Lomond Reservoir, the change in the storage capacity of the reservoir was estimated. Cross sections from the previous surveys, along with the previously determined reservoir bed surfaces, were compared to determine the magnitude of sedimentation throughout the reservoir. Results of the 1998 investigation, which were determined from changes in the cross-sectional areas along transects, indicated that the capacity of the reservoir decreased 55 acre-ft, less than 1 percent of total capacity, from 1982 to 1998 (McPherson and Harmon, 2000). To determine the most recent (1998–2009) storage capacity changes in the reservoir, bathymetric and topographic surveys of the reservoir were conducted. The bathymetric survey maps the depth to the measurable wetted reservoir bed

below the water surface, within the instrument capabilities, whereas the topographic survey maps the reservoir-bed surface above the existing water line to the point at which vegetation is too dense to determine the ground surface. Addition of this topographic survey to the bathymetric survey provides information about the changes in the upper reach of the reservoir where water is shallow or the reservoir is dry, as well as shoreline changes throughout the reservoir. Bottom material samples were collected throughout the reservoir to determine the location of sedimentation and to infer the sedimentation processes operating in the reservoir by examining the physical characteristics of the bottom material. By defining the change in capacity and the extent and nature of the sedimentation, the results of the investigation provide an accurate water storage capacity at a given water-surface altitude that water managers can use to determine current water supply availability and evaluate changes in water storage capacity to make effective water-management decisions.

The USGS and the CSUMB SFML have a mutual interest in developing an understanding of anthropogenic factors and climatological effects impacting the capacity of, and management practices associated with, California's reservoir-based water supplies. The long-term effects of climate change on water-resource management are of great concern to State and local water managers. The techniques employed in this study to improve understanding of the quantitative effects of increased sedimentation rates may allow for a more effective assessment of changes in water-storage capacity in other, similar basins. Knowledge of the potential reductions in storage capacity due to sedimentation may also help water managers more effectively adjust storage dynamically to prevent flooding. With this study, the USGS has a unique opportunity to document advanced methods used for bathymetric surveys of small reservoirs occupying steep and narrow drainages in mountainous terrain.

Purpose and Scope

The purpose of this report is to (1) document the use of advanced instrument technology to complete new bathymetric and topographic surveys of Loch Lomond Reservoir; (2) document data-processing techniques; (3) present an estimate of the storage capacity of the reservoir in 2009; and, (4) present a comparison of the survey results with those of the 1998 investigation and the resulting determination of the change in storage capacity since 1998, done to improve understanding of the dynamics of Loch Lomond Reservoir.

The thalweg, the deepest continuous channel along the reservoir valley floor, and selected transects were mapped and are illustrated to show the variations in the methods used in the current and previous investigations and to document locations of sediment deposition and erosion. Variations in the methods used for these investigations limit the number of data points available to plot each of the cross-sectional areas. A revised bathymetric map, a stage-capacity curve for storage

Figure 2. Completed earth-fill Newell Creek Dam and Loch Lomond Reservoir, Santa Cruz County, California, about 1961. Photograph provided by Chris Berry, Water District of the city of Santa Cruz, California.

capacity of the reservoir at multiple stages (2-ft intervals), and a stage-surface-area curve at the same intervals are provided in this report.

The particle-size distribution of the sediments composing the reservoir bed material at selected locations was documented and compared with that measured during previous investigations. This information will help determine the sedimentation processes and the extent of the sedimentation in Loch Lomond Reservoir independent of the effects of changes in storage capacity.

Description of Study Area

Loch Lomond Reservoir, located 9 mi north of Santa Cruz, is 2.5 mi long with a maximum width of about 1,500 ft. Newell Creek starts in the Santa Cruz Mountains, approximately 3 mi above the upper end of the reservoir, and flows into the San Lorenzo River 2 mi downstream from the dam

(*fig. 3*). The Newell Creek watershed contributing area above Loch Lomond is approximately 8.25 mi². The contributing drainage area extends from the top of the Santa Cruz Mountains, at an altitude of more than 2,300 ft, to the spillway, at an altitude of 577.5 ft, and is underlain predominantly by interbedded layers of sandstone, siltstone, and shale of Tertiary age. These materials decompose into soil that is easily eroded and can be prone to landslides (Brown, 1973).

The climate in Santa Cruz County is Mediterranean, with wet, mild winters and dry, warm summers. Annual rainfall in Santa Cruz averages about 30 in., with an average of 25 in. from December through February (California Data Exchange Center, 2009). Rainfall often arrives in large storms producing substantial runoff with daily mean discharge far exceeding mean annual streamflow values, which increases the probability of sediment transport in natural river channels (Heimann, 2001). *Figure 4* shows that annual peak streamflow at a gaging station near Loch Lomond Reservoir is periodically large (greater than 10,000 ft³/s [cubic feet per second]), and could transport a large amount of sediment to the reservoir.

Methods

Several methods have been used to monitor the storage capacity of, and the rate of sedimentation in, Loch Lomond Reservoir over the years of its operation. Each of these methods requires different equipment for data collection and different techniques for data processing, and therefore has a different accuracy level. In an effort to accurately define the storage capacity of the reservoir, the USGS examined each of these methods to determine the most accurate and cost-effective approach for performing bathymetric and topographic surveys of the reservoir bed. In the sections below, each of these methods and its limitations are documented, and the application of a new, state-of-the-art method for combined bathymetric and topographic surveying to establish a new baseline for calculations of reservoir stage capacity is described. Finally, to supplement the bathymetry survey and to enhance understanding of sedimentation processes in Loch Lomond Reservoir, the sampling and analysis of the reservoir bed sediment to determine grain size are discussed.

Bathymetry

As mentioned previously, bathymetry is the measurement, within the instrument capabilities, of the depth of the wetted reservoir bed below the water surface. These surveys are often supplemented with some type of topographic survey to obtain land-surface altitude data, above those determined from the bathymetric survey to the spillway crest altitude or higher. As instrumentation and data-processing methods have improved, the survey method used at Loch Lomond Reservoir has changed.

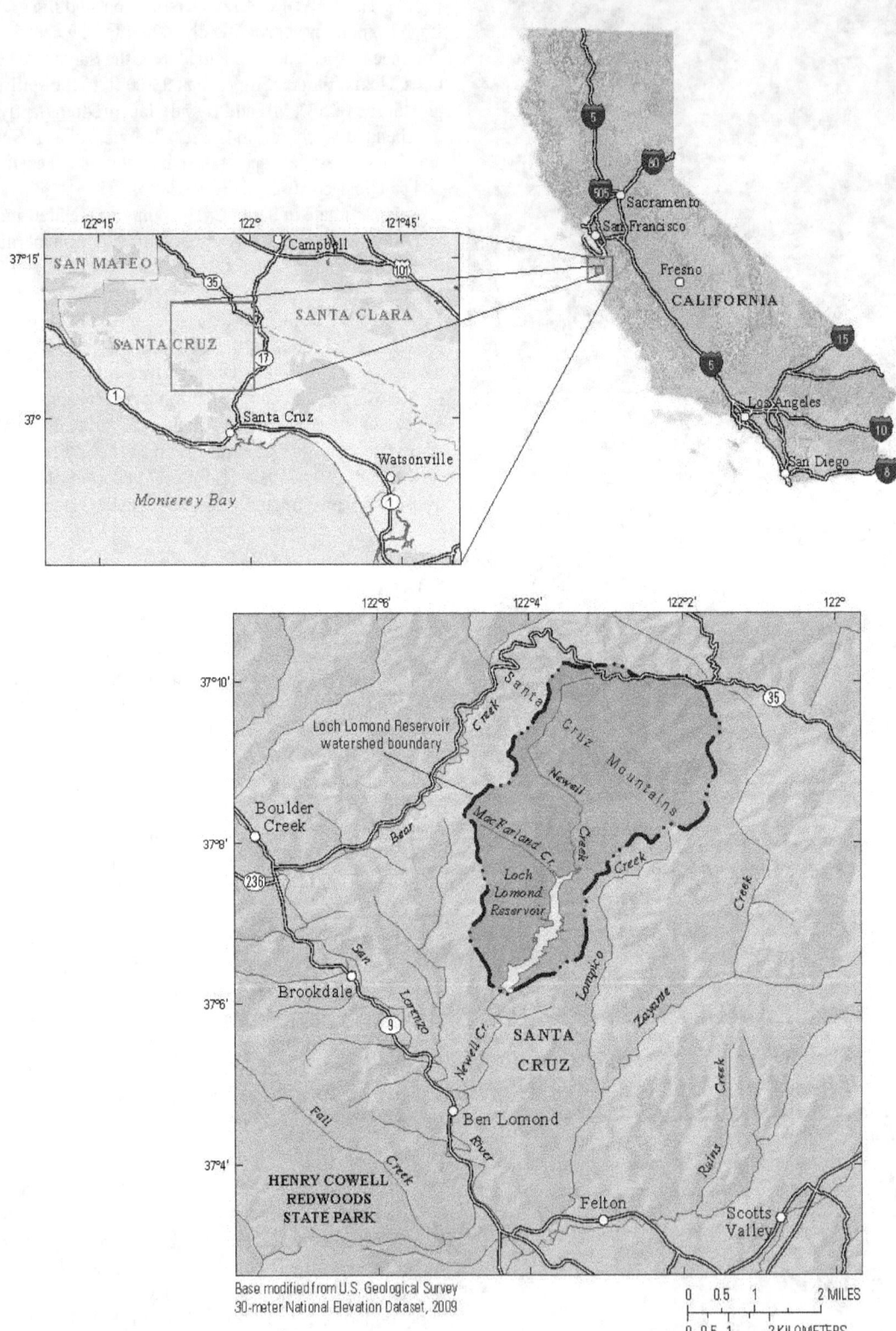

Figure 3. Location of Loch Lomond Reservoir in the Santa Cruz Mountains, contributing watershed, and related features, Santa Cruz County, California

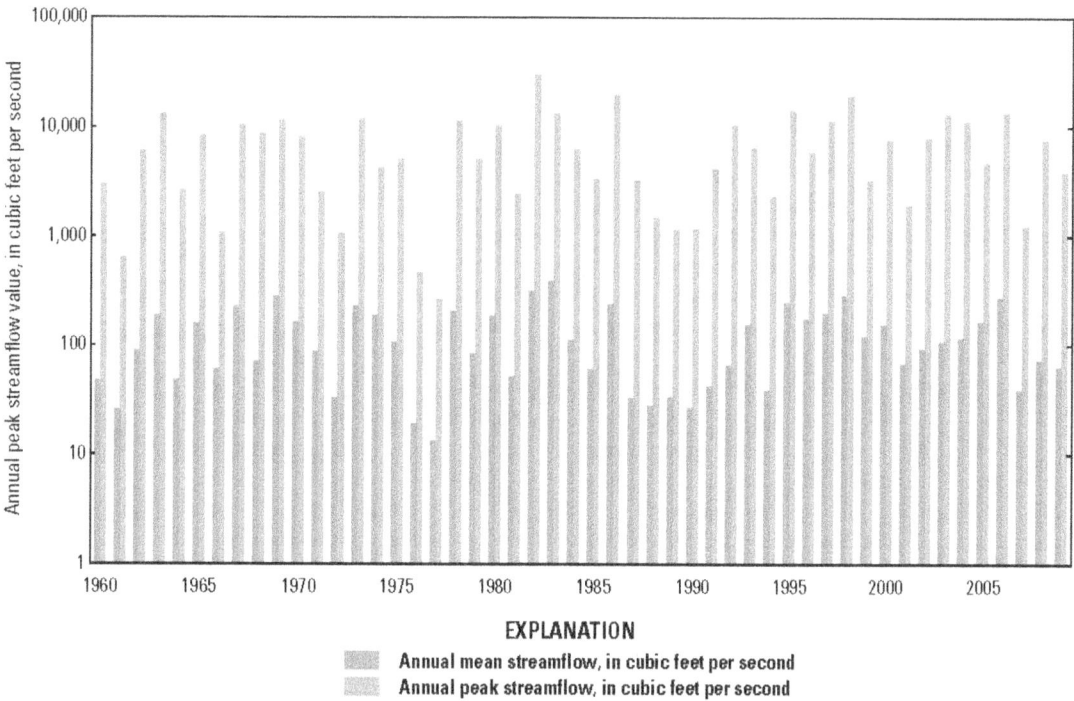

Figure 4. Annual peak streamflow and annual mean streamflow at U.S. Geological Survey streamgaging station 11160500, San Lorenzo River at Big Trees, California, upstream of Loch Lomond Reservoir.

Previous Investigations and Limitations

Historically, several surveying methods have been used to measure the surface area of the bed of Loch Lomond Reservoir. In addition, many data-processing techniques have been used to calculate the volume of the reservoir, making comparisons between years difficult or even impossible where data are very limited. These previously used surveying methods and data-processing techniques were investigated to determine the levels of accuracy that could be expected. These previously used methods are not well documented, which limits their usefulness; as new technologies and methods are developed, documentation is needed in order to be able to reproduce the results for future investigations. Previously used data-collection methods include photogrammetry, tag lines and fathometers, and global positioning system (GPS)-enhanced echo sounders. Survey methods were evaluated for vertical and horizontal accuracy, and an assessment was done to determine how well each survey was spatially distributed throughout the reservoir.

1960 Pre-Dam Construction

Photogrammetry was used to define contour lines, or lines of equal altitude, in 1960, and is typically used to map the topography of a large area. Ground-truth surveys are typically performed in conjunction with photogrammetry to tie known locations and altitudes to the photogrammetry data. The control points established by the ground-truth surveys were accurate to ±0.1 ft and were referenced to a local or mean-sea-level datum. The average vertical accuracy of the photogrammetry, in general, is expected to be about 1.2 ft with a 95-percent confidence level (Federal Emergency Management Agency, 2006).

The 1960 storage capacity of the reservoir was estimated to be about 8,600 acre-ft. Because this was a rough estimate, an improved estimate of the potential storage capacity at that time was obtained by using the contour lines to define the reservoir-bed surface. This method allowed data-processing techniques to be similar to those used in the investigations completed in 1982 and later. Because the contour lines were not available from a digital source, the USGS scanned a 10-ft-interval topographic contour map produced from this survey, registered the rasterized map image with known coordinates, and converted the map image into vectorized contours, which were then attributed with the altitude values from the map. Because none of the dam structures are in place on the pre-1960 map, estimated contours of the structures from the 1998 survey were used to place the estimated contours for pre-1960 at the dam. These digital contour lines were used to estimate the original capacity at the time the dam was first completed with the approximate dam structure. This data-processing method for determining capacity is similar to the method used in the 1998 investigation. Contour values and the thalweg were used to guide the interpolation of the area between contour lines. Additionally, where data were lacking, known spot altitudes were used.

1971 Investigation

The primary goal of the 1971 investigation was to determine the amount of sediment that had been deposited over a 2 year period in the Loch Lomond Reservoir, and the potential rate of future sedimentation. Suspended sediment was monitored at the inflow and outflow during the 1970 and 1971 water years (Brown, 1973), and the volume of sediment already deposited was estimated. Bathymetric data were collected by surveying along transects shown in *figure 5*. These transects and additional spot altitudes were used to create a new bathymetric map of the reservoir; however, storage capacity was not estimated. Limited data from this and the previous survey did not provide enough information to accurately estimate the total storage capacity. As indicated by Brown (1973), the 1960 and 1971 reservoir-bed surfaces could not be compared; however, by using the transects to calculate change in storage capacity since 1960, the volume of sediment that had accumulated during 1960–71 was estimated to be at least 46 acre-ft.

To compare the limited 1971 bathymetric survey data with data from the 1960 pre-construction investigation, a 1971 topographic map of Loch Lomond Reservoir (Brown, 1973) was scanned, registered with known coordinates, vectorized, and attributed in the same way as the pre-1960 map. This map had less topographic detail than the pre-1960 map as a result of the surveying method used at the time. Comparison of the reservoir-bed surface mapped in the 1971 investigation with that indicated in the pre-1960 map shows substantial changes in the contours in the geographical areas in-between the transects where no new measurements had been made.

Because of the limitations of the survey equipment and data-processing methods available at the time of the 1971 investigation, the inability to compare the surface area to the previously defined surface area precludes a comparison of storage capacity. Instead, transects from this investigation were examined to estimate changes in the bed surface throughout the reservoir. This method gives only an approximation of the changes throughout the reservoir bed.

Because depth is measured only along the transects and at important reservoir-bed surface features, accuracy for this type of survey is difficult to determine—that is, although the accuracy of the vertical measurements is good, the spatial distribution of the data points used to make a capacity calculation could be so poor that it is not reasonable to make a storage capacity estimate. As mentioned previously, the combination of aerial photogrammetry and depth soundings or fathometer readings provides a vertical resolution of about 1.2 ft. If a data point is assumed to represent an area with a 1-ft radius, the bathymetric survey data represent only about 0.5 percent of the total reservoir water surface area, which means that spatial distribution of data is so poor that storage capacity could not be determined (Brown, 1973).

1982 Investigation

In 1982, permanent ranges, or transects, were established and intended for use in future studies. The 1982 bathymetric survey was similar to the 1971 bathymetric survey in that depth measurements were made along transects. A fathometer was used to determine depth. The transects were then mapped using bearings and distances. Because the reservoir is generally long and narrow and visibility is limited by the steep canyon walls, this type of survey has the potential to become increasingly horizontally skewed from one end of the lake to the other. The fathometer typically has a resolution of about 0.1 ft, with a minimum error of ±0.25 percent (International Hydrographic Office, 1998); by using a tag line or an incremented rope between each of the transect end points, a measurement could be made within 1 ft horizontally of the surface-measurement point. It is not known what the horizontal accuracy along the tag lines would have been because there are no known corrections for boat movement at the time of the survey.

Comparison of thalweg transects indicated that deposition had occurred in the lower reach of the reservoir as a result of landsliding and in the upper reach as a result of sediment inflow from Newell Creek (Fogelman and Johnson, 1985). The total storage capacity was computed, by using a method described by Eakin (1936; revised by Brown, 1939), to be 8,824 acre-ft. This data processing method uses the surface area at 10-ft contours to calculate the volume of each prismoid, as described in the 1982 investigation by Fogelman and Johnson (1985), who also describe the inaccuracy of the base maps and initial surveys, which prohibited comparison with the 1971 computations; therefore, an estimate of the volume of sedimentation was not attempted.

In 1998, the contoured altitude data collected in 1982 were digitized into a geographic information system (GIS) to be compared with those in the 1998 investigation. As described in McPherson and Harmon (2000), information associated with the 1982 depth measurements for locating the altitude data spatially was limited. In an effort to better horizontally locate the altitude data obtained from the August 1982 survey, the digital contour lines and transects were relocated by using shoreline features obtained from the 2009 survey that were unlikely to have eroded or moved. Several methods of transformation were examined to place the map features without having too great an effect on the spatial relations of the contours. An affine, or linear, transformation to place the entire data set was not possible. Instead, a rubber sheet transformation was required to place the transects and contours in the proper location. Features between transects were often missed, causing calculations of total capacity to be inaccurate.

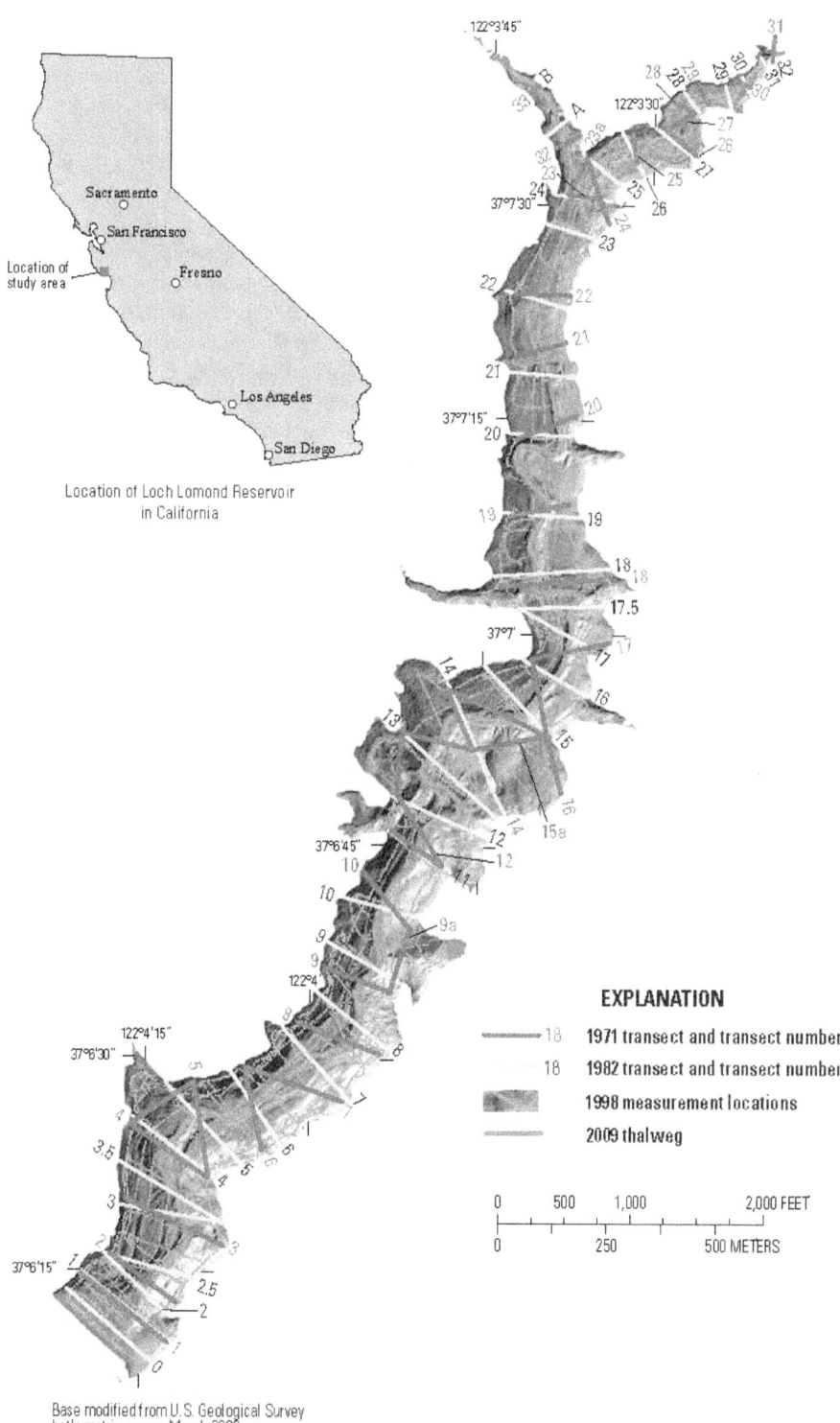

Figure 5. Locations of 1971 and 1982 transects, 1998 measurement locations, and the 2009 thalweg, Loch Lomond Reservoir, Santa Cruz County, California.

1998 Investigation

In an effort to reduce the uncertainty from missing altitude data between transect lines surveyed in the previous investigations, the USGS surveyed the reservoir bed by using a combination of GPS and electronic depth sounding equipment mounted on a boat. As described in McPherson and Harmon (2000), the 1998 survey had a 0.7-m (2.3-ft) horizontal resolution and a 0.002-m (0.01-ft) vertical resolution. This method proved to be superior to previously used methods for collecting depth data throughout the reservoir and capturing features along pre-existing transects as well as between the transects. Time constraints and the level of technology of the equipment available did not allow for a detailed shoreline survey, so aerial photography showing the surface-water boundary and a 30-m (98.4-ft) digital elevation model (DEM) of the shoreline terrain were used to estimate the shoreline altitudes. Reservoir bed altitudes were interpolated from the water surface to the closest available data collection location from the boat survey. Important advantages to using this method are that there are no transect posts that need to be maintained, and the altitude measurement locations are well defined with GPS coordinates. This type of survey is easily completed in 2 days for Loch Lomond Reservoir, which means that it could be completed under ideal weather and lake conditions.

Although this survey method records features that may have been missed by the previous transect methods, some limitations still exist. The boat-mounted equipment is capable of recording values up to as shallow as only 2 ft below the water surface. Therefore, the survey was completed when the reservoir level was only 4 ft below the spillway altitude of 577.5 ft, approximately 6 ft of the shoreline surface could not be defined. The shoreline contour at 577.5 ft was defined by using aerial photography with an approximate horizontal resolution of about 1 m (3.28 ft). Use of this method to define the shoreline is dependent on the absence of overhanging trees and can provide only an approximate location for the spillway-altitude contour. Surveying the shoreline would allow for a more accurate determination of the maximum storage capacity.

In addition to shoreline-data limitations, only about 3.4 percent of the total surface area, at a 1-m (3.28-ft) resolution, is represented with this type of survey. As in the 1971 and 1982 investigations, in which only transects were surveyed, unobserved features are present that may add to or subtract from the storage capacity of the reservoir (*fig. 6*).

Interferometric Bathymetric Sidescan Sonar and LiDAR Topographic Survey

In an effort to conduct an investigation that would use repeatable surveying techniques and produce highly accurate and replicable results, the USGS was assisted by the CSUMB SFML research team, who provided instrumentation and data processing expertise in the 2009 investigation. The state-of-the-art combination of a bathymetry survey and a topographic survey was performed in late March 2009, when the reservoir was at an optimum level (577.5 ft) to allow boat access throughout the lake. Bathymetry of the reservoir bed was surveyed by using boat-mounted SEA SwathPlusH interferometric bathymetric multibeam-sidescan sonar, and topographic data were collected with a boat-mounted mobile laser scanning system (Light Detection and Ranging, or LiDAR). The LiDAR survey provided highly detailed land-surface coverage of the areas seasonally exposed during dry periods, thereby producing a more accurate bathymetric map and storage-capacity table that could otherwise be obtained only when the reservoir is at or near maximum capacity. This laser scanner is capable of achieving decimeter accuracy with sub-meter resolution at up to a 1-km range (California State University Monterey Bay Seafloor Mapping Lab, 2010). The R/V Mac-Ginitie, CSUMB's research vessel, was used as the platform for both the bathymetric and topographic surveys (*fig. 7*). Final altitude and coordinate data were analyzed by CSUMB and provided to USGS. Three redundant position controls were used during the surveys: (1) A GPS reference station placed over a previously established benchmark on the dam crest was used as the master geodetic horizontal and vertical control for the project, (2) another previously established benchmark on the spillway retaining wall was used as a check on vertical control for the laser topographic results, and (3) the reservoir water-level staff values were recorded at the beginning of the surveys as another accuracy check on vertical control. The advanced technology data collection using an interferometric bathymetric sidescan-sonar survey and a LiDAR topographic survey was tested and verified by using a number of quality-assurance techniques. These methods were documented by CSUMB SML and are presented in *appendix A* of this report. Detailed descriptions of control points, photos, and pertinent quality-control checks are also provided. The altitude of the reservoir-bed surface resulting from this detailed bathymetric and topographic survey was used to calculate the March 2009 storage capacity of Loch Lomond Reservoir.

Storage Capacity

According to Vanoni (2006), "The average end area method is better suited for application to reservoirs having fairly uniform width throughout its length and ranges are established normal to the stream thalweg as possible." This statement was found to be true in the analysis of the 1998 and 2009 investigation results. A comparison of the volumes calculated for 1998 and 2009 by using this method showed a loss of storage capacity of about 25.5 acre-ft. This average end area method assumes that the volume between two consecutive transects, or end areas, is the average of their areas multiplied by the distance between them. This method, however, does not represent the diverse features found in Loch Lomond Reservoir because the characteristics of the reservoir-bed surface between the transects are not considered.

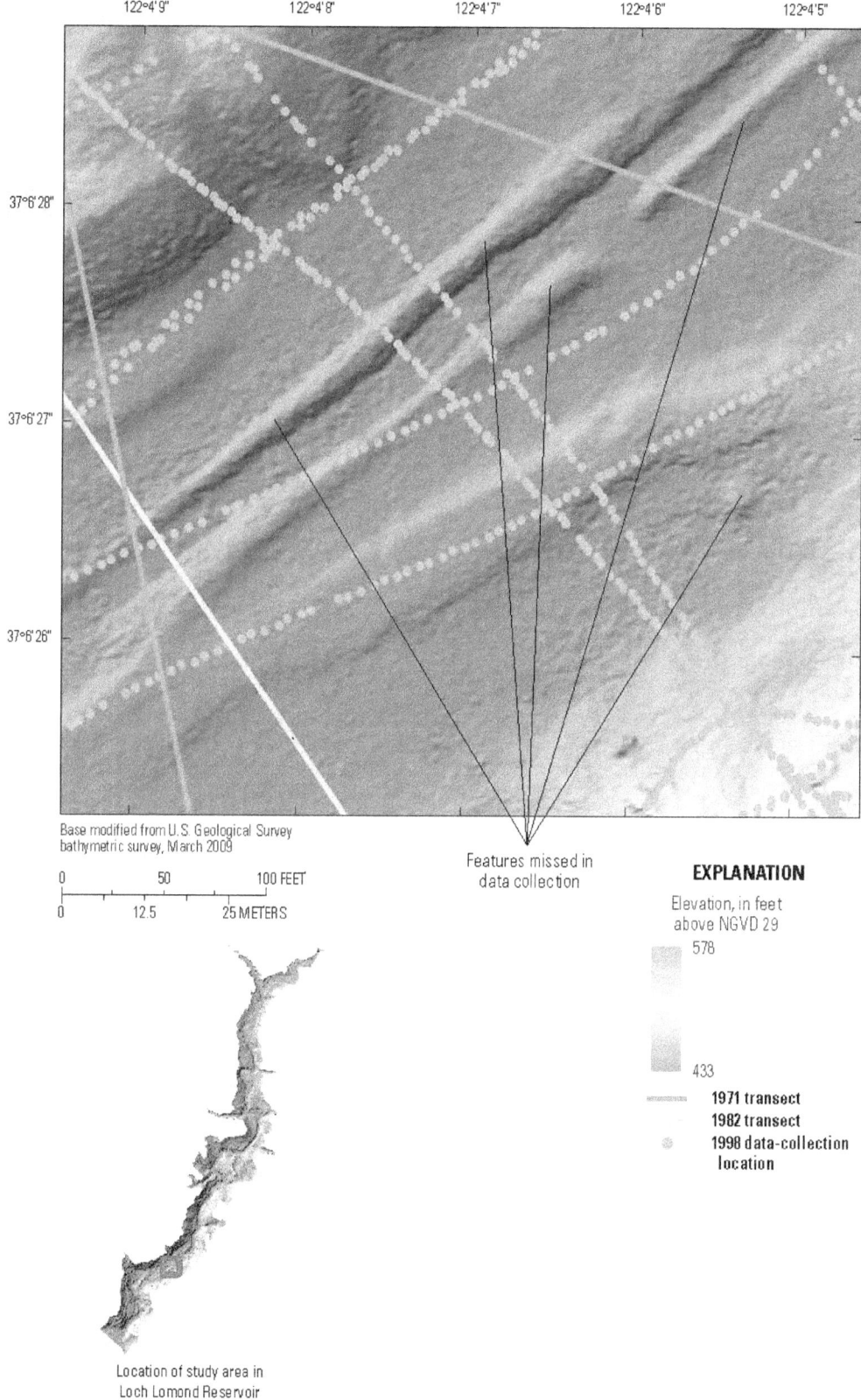

Base modified from U.S. Geological Survey
bathymetric survey, March 2009

Features missed in
data collection

Location of study area in
Loch Lomond Reservoir

EXPLANATION

Elevation, in feet
above NGVD 29

578

433

1971 transect
1982 transect
1998 data-collection
location

Figure 6.   Bathymetric features that were missed with 1971 and 1982 survey data-collection methods, thus altering storage capacity calculations for Loch Lomond Reservoir, Santa Cruz County, California.

Figure 7.   SEA SwathPlusH 468-kHz interferometric bathymetric sidescan-sonar system pole-mounted on the bow of the R/V MacGinitie, tied up at floating docks in front of the launch ramp and parking area, Loch Lomond Reservoir, Santa Cruz County, California. Photograph provided by Rikk Kvitek, California State University, Monterey Bay, Sea Floor Mapping Laboratory.

To accurately measure the storage capacity, all available data are needed to represent the reservoir-bed surface as accurately as possible. Loch Lomond has a steep and rough reservoir-bed surface with many features that could add or subtract substantial volumes to or from the total storage capacity. A greater amount of data spatially distributed throughout the reservoir-bed surface was observed in 1998 by using a depth echo sounder enhanced with GPS (McPherson and Harmon, 2000). This previous investigation utilized 28,000 data points throughout the reservoir-bed surface, along with an estimated water-surface contour to create an altitude model of the reservoir surface.

To most accurately represent the reservoir-bed surface using the available data, a triangulated irregular network (TIN) model was created by using ArcInfo geographic information system (GIS) software (ESRI, 2008). A TIN model is a surface representation derived from irregularly spaced points with an x, y coordinate and a z value or surface value. The ArcInfo GIS software has the ability to quickly perform volumetric calculations from these surface models. Given a set of points, many possible triangulations can be created. ArcInfo uses the Delaunay triangulation algorithm to optimize the surface model. This algorithm creates triangles that collectively are as close to the equilateral shapes as possible. This method of interpolation keeps altitudes at new points as near as possible to known input points. These TIN models were used to create surface areas for each of the previous and the most current investigations, which were then used for storage-capacity calculations and comparisons.

During the 2009 bathymetric and topographic survey, more than 13 million data points were measured and used in the TIN model. This procedure resulted in an extremely complex surface model with a fine resolution from which to determine the storage capacity from the reservoir-bed surface up to the spillway altitude.

Data Collection

During March 28–30, 2009, the USGS and CSUMB SFML surveyed Loch Lomond Reservoir using a combination of mobile, vessel-mounted topographic LiDAR (terrestrial laser scanner) and interferometric bathymetric sidescan sonar for full-basin bare-earth mapping to determine storage capacity.

The interferometric sidescan-sonar system was able to map bathymetry from the reservoir floor to the water surface, and the mobile laser scanner covered the exposed basin topography from the water surface to the top of the spillway retaining wall. Manual and automated data cleaning and filtering removed the effects of all water-column debris and terrestrial vegetation, yielding a high-density, merged bathymetric/topographic xyz point cloud containing more than 130 million individual data points. These points were used to create final cleaned bare-earth gridded xyz data sets of the entire reservoir basin at 0.5-m (1.64-ft) resolution containing 30,634,431 points that met or exceeded IHO Special Order standards for hydrographic surveys (International Hydrographic Office, 1998).

Details of this method are presented in *appendix A* of this report, which also documents quality-assurance methods and other steps taken to ensure high standards for these hydrographic surveys.

Storage Capacity Calculations

By using the modeled reservoir-bed surface from the 2009 bathymetric and topographic survey data, storage capacity of Loch Lomond Reservoir was calculated for 2-ft water-surface altitude intervals (stage) up to the maximum water storage capacity of the reservoir (a stage of 577.5 ft) (*table 1*). The estimated maximum capacity of the reservoir in March 2009 was 8,646 acre-ft.

A plot of the relation between water storage capacity at 2-ft stage intervals to the maximum capacity of the reservoir and water-surface area (*fig. 8*) illustrates that capacity has been reduced fairly consistently, with the greatest reduction at the high altitudes and zero reduction at the bottom of the reservoir. The plot of water-surface area and water-surface stage

(altitude) shows that some erosion or reservoir-bed changes may have occurred from 577.5 to about 566 ft. These changes could be an indication of wind wave erosion, erosion near the inlet of Newell Creek, or simply the presence of features that were not mapped adequately in the 1998 investigation. This curve also shows that the water-surface area decreased most from the bottom of the reservoir to an altitude of about 470 ft, likely indicating where sedimentation has been greatest.

Sedimentation

Watershed history, including history of climate, fire, land-use change, earthquakes, forest logging, and road construction, plays a key role in determining the volume of sedimentation expected. Measurements of sedimentation by means of sequential reservoir capacity surveys were completed in 1971, 1982, and 1998. These surveys provided general sedimentation rates with limited success as a result of the differences in methods and precision among the surveys (Brown, 1973; Fogelman and Johnson, 1985; McPherson and Harmon, 2000).

Table 1. Storage capacity for water-surface altitudes (stage) at 2-foot intervals to the maximum capacity of Loch Lomond Reservoir, Santa Cruz County, California, March 2009.

[All altitudes are in feet above NGVD 29]

Water-surface altitude	Storage capacity, in acre-feet	Water-surface altitude	Storage capacity, in acre-feet	Water-surface altitude	Storage capacity, in acre-feet	Water-surface altitude	Storage capacity, in acre-feet
[1]577.51	8,646	542	3,810	506	1,262	470	141
576	8,386	540	3,620	504	1,169	468	113
574	8,046	538	3,436	502	1,081	466	92
572	7,711	536	3,259	500	996	464	77
570	7,384	534	3,088	498	915	462	65
568	7,067	532	2,923	496	837	460	54
566	6,762	530	2,764	494	763	458	45
564	6,467	528	2,611	492	692	456	37
562	6,180	526	2,462	490	625	454	31
560	5,902	524	2,319	488	562	452	26
558	5,634	522	2,180	486	501	450	21
556	5,375	520	2,048	484	445	448	17
554	5,125	518	1,920	482	392	446	13
552	4,883	516	1,798	480	341	444	10
550	4,651	514	1,681	478	294	442	7
548	4,428	512	1,570	476	250	440	5
546	4,213	510	1,462	474	210	438	3
544	4,008	508	1,360	472	173	436	1

[1]Maximum capacity.

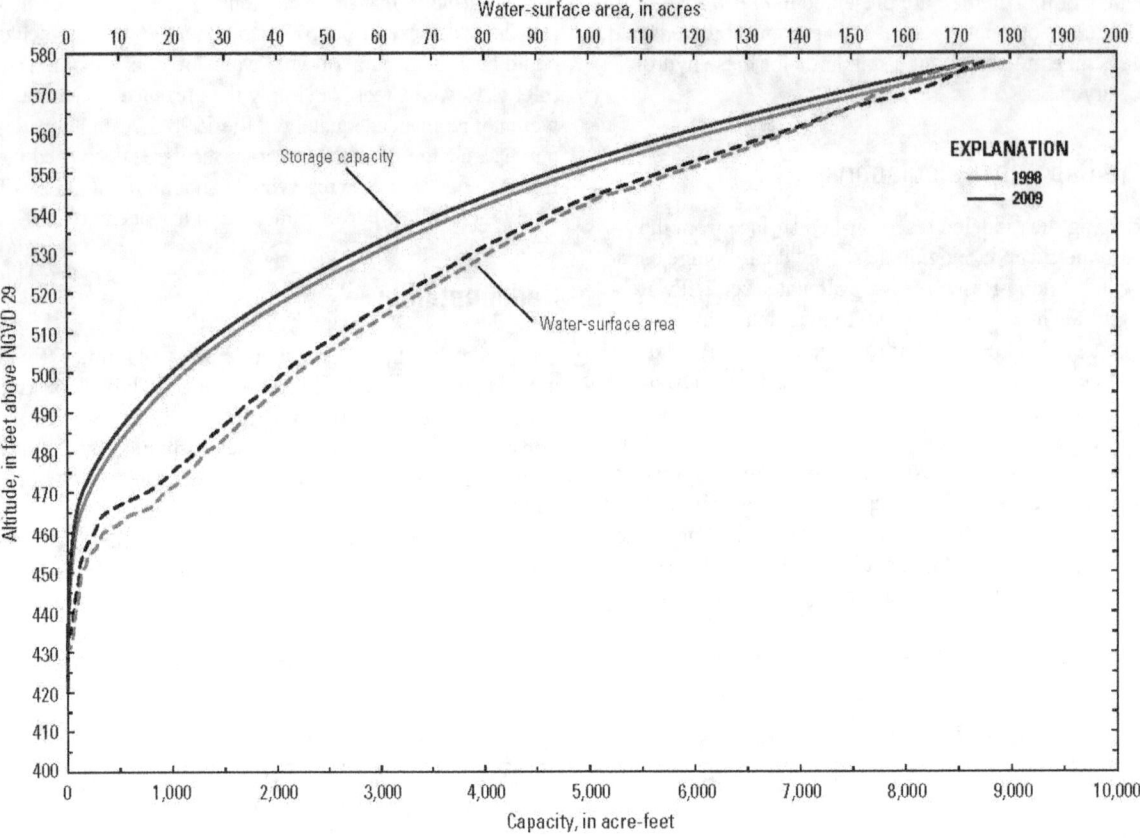

Figure 8.   Water-surface area and stage-capacity curves for Loch Lomond Reservoir, California, November 1998 and March 2009.

Data Collection

Bottom-material sediment samples were collected as part of the 2009 investigation to supplement the previous sediment surveys at points representing the reservoir thalweg location. The 2009 samples were collected as close as possible to the points identified during these previous surveys. The water surface was at a stage of 577.15 ft during sample collection, affording navigable access by boat to nearly the entire reservoir. Samples were collected at selected points along the reservoir thalweg at the intersection with selected 1982 transects, and in the upper portion of the reservoir (delta formation area) at selected intervals along the 1982 transects (*fig. 9*). Samples were collected from a boat furnished and operated by the Loch Lomond Recreation Area Park Ranger using a US BMH-60 sediment sampling instrument (*fig. 10*) deployed using a USGS B-56 type sounding reel.

The US BMH-60 sediment sampler instrument is a hand-line bed-material sampler that penetrates approximately 1.7 in. into the bed material. This lightweight aluminum sampler is 22 in. long, weighs 32 lbs, and is designed to be suspended from a flexible line and lowered and raised directly by hand or by use of a hand-powered reel. Ballast makes the sampler nose heavy by about 4 lbs to help the sampling bucket penetrate the bed material of the reservoir. The sampling bucket can hold approximately 10.5 in^3 (175 cm^3) of material and is spring loaded by cross-curved, constant-torque, motor-type springs. Tension on the suspension line allows the bucket to be cocked in the open position by means of a wrench. Once the bucket is fully retracted within the body shell of the sampler, it is ready to collect a sample. When the tension on the suspension line is reduced to a specified amount, the spring-loaded cocking device releases the bucket mechanism. The rapidly closing bucket penetrates the bed surface and completely encloses a sample of the bed material. The sampler is then raised to the surface and the bed material sample is transferred to a container. The container is sealed and refrigerated to minimize the possibility of modifying the sample prior to lab analysis.

As a result of the limited cable length of the B-56 sounding reel, an additional length of rope was added to the end at the sampler so that locations where depth exceeded 140 ft could be sampled. The reel system was mounted temporarily to the bow of the boat (*fig. 11*). Where depths were shallow, the sampler was lowered and raised by hand using a rope hand line. Particle-size distributions of samples collected in 1971, 1982, 1998, and 2009 are listed in *appendix B* of this report.

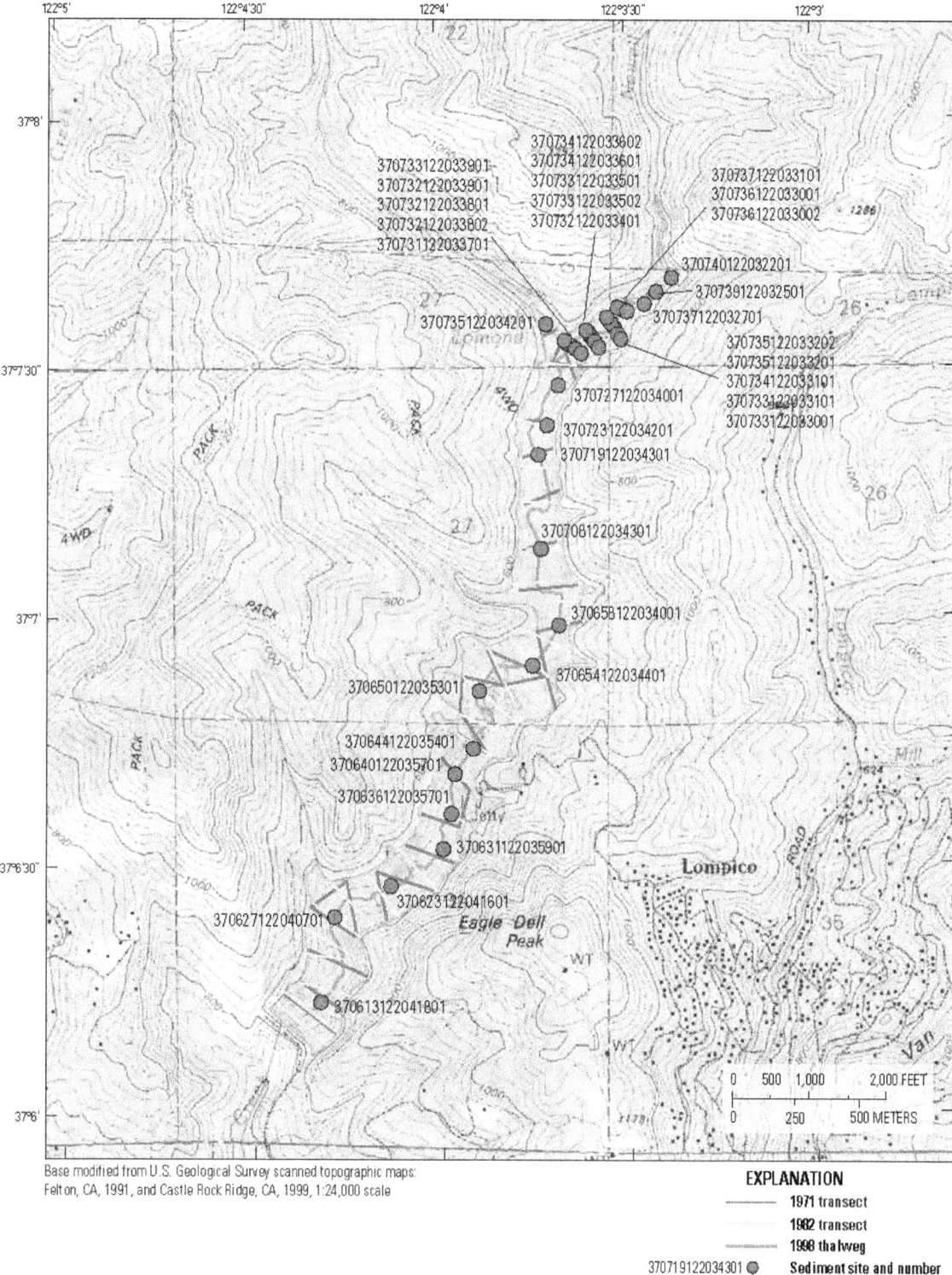

Figure 9. Reservoir bed sediment sampling locations, May 19, 2009 at Loch Lomond Reservoir, Santa Cruz County, California.

Figure 10. Photograph of a United States series bed-material hand-line sediment sampler instrument developed in 1960 (US BMH-60).

Sampling stations were selected to coincide with the 22 stations at which samples were collected in the 1982 study (*fig. 9; table 2*). Samples also were collected at 14 additional stations at several upstream cross sections on either side of the thalweg stations to more accurately define the sedimentation pattern in the Newell Creek alluvial fan area, resulting in a total of 36 pre-selected sampling locations. Coordinates of these sampling stations were determined by mapping previous transects with the thalweg location of that particular investigation. All sediment data are archived in the USGS National Water Information System (NWIS) water-quality database and are available to the public (*http://waterdata.usgs.gov/ca/ nwis*). Individual NWIS station identifiers were assigned and established in the NWIS site file by using these coordinates. The sampling stations were located in the field during sampling by using a handheld GPS unit with a horizontal accuracy of approximately 5 ft. This horizontal accuracy coupled with movement of the boat caused by wind during the sampler deployment at each station resulted in a location error of no more than 10 ft.

The particle-size distribution of reservoir bottom sediment samples was determined by a combination of wet-sieve and pipette analysis methods (Guy, 1969). The samples were analyzed at the USGS California Water Science Center (CAWSC) Sediment Lab in Marina, California. The particle-size data were used to determine general changes in the size of the sediment that was deposited in the reservoir, particularly in the Newell Creek alluvial fan area, between the previous investigations and 2009.

Reservoir Transects

Individual transects were examined by using the available altitude data from each investigation year to determine changes in the reservoir bed surface over time. Because transects were used in the 1971 and 1982 investigations, data from these transect locations were compared with the corresponding data locations from those investigations that were independent of transect surveys.

Because previous bathymetric methods did not capture all the features of the reservoir bed, the quality of each transect in each investigation was evaluated; altitude data that appeared to be poorly mapped were not used. Where possible, these transects were compared to determine the percent change of cross-sectional area from one investigation to the next. Changes in reservoir bed altitudes are shown in *appendix C* of this report, which includes all of the transect profiles measured (*fig. 5*).

Figure 11. Boat and reel system on crane platform for sediment sampler

Table 2. Sediment sampling stations used for the 2009 investigation, Loch Lomond Reservoir, Santa Cruz County, California.

[USGS, U.S. Geological Survey. Latitudes and longitudes are shown in degrees, minutes, and seconds. **Abbreviation:** Lat, latitude; Long, longitude]

USGS station number	Station name	Transect number (see fig. 5)	Location (NAD 83)	
370613122041801	Loch Lomond Reservoir Thalweg A Range 2 near Ben Lomand	2	Lat: 37° 06′13.5″	Long: −122°04′18.7″
370623122041601	Loch Lomond Reservoir Thalweg A Range 4 near Ben Lomand	4	Lat: 37 06′23.7″	Long: −122 04′16.4″
370627122040701	Loch Lomond Reservoir Thalweg A Range 6 near Ben Lomand	6	Lat: 37 06′27.5″	Long: −122 04′07.5″
370631122035901	Loch Lomond Reservoir Thalweg A Range 8 near Ben Lomand	8	Lat: 37 06′ 31.9″	Long: −122 03′59.2″
370636122035701	Loch Lomond Reservoir Thalweg A Range 9 near Ben Lomand	9	Lat: 37 06′36.2″	Long: −122 03′57.8″
370640122035701	Loch Lomond Reservoir Thalweg A Range 10 near Ben Lomand	10	Lat: 37 06′40.9″	Long: −122 03′57.3″
370644122035401	Loch Lomond Reservoir Thalweg A Range 11 near Ben Lomand	11	Lat: 37 06′44.0″	Long: −122 03′54.4″
370650122035301	Loch Lomond Reservoir Thalweg A Range 13 near Ben Lomand	13	Lat: 37 06′51.0″	Long: −122 03′53.4″
370654122034401	Loch Lomond Reservoir Thalweg A Range 15 near Ben Lomand	15	Lat: 37 06′54.1″	Long: −122 03′44.8″
370658122034001	Loch Lomond Reservoir Thalweg A Range 17 near Ben Lomand	17	Lat: 37 06′58.8″	Long: −122 03′40.8″
370708122034301	Loch Lomond Reservoir Thalweg A Range 19 near Ben Lomand	19	Lat: 37 07′08.1″	Long: −122 03′43.5″
370719122034301	Loch Lomond Reservoir Thalweg A Range 21 near Ben Lomand	21	Lat: 37 07′19.4″	Long: −122 03′43.9″
370723122034201	Loch Lomond Reservoir Thalweg A Range 22 near Ben Lomand	22	Lat: 37 07′23.1″	Long: −122 03′42.6″
370727122034001	Loch Lomond Reservoir Thalweg A Range 23 near Ben Lomand	23	Lat: 37 07′27.8″	Long: −122 03′40.7″
370732122033801	Loch Lomond Reservoir Thalweg A Range 25L2 near Ben Lomand	25	Lat: 37 07′32.6″	Long: −122 03′38.8″
370732122033901	Loch Lomond Reservoir Thalweg A Range 26L2 near Ben Lomand	25R1	Lat: 37 07′32.9″	Long: −122 03′39.3″
370733122033901	Loch Lomond Reservoir Thalweg A Range 25 near Ben Lomand	25R2	Lat: 37 07′33.2″	Long: −122 03′39.7″
370732122033802	Loch Lomond Reservoir Thalweg A Range 25L1 near Ben Lomand	25L1	Lat: 37 07′32.2″	Long: −122 03′38.2″
370731122033701	Loch Lomond Reservoir Thalweg A Range 25R1 near Ben Lomand	25L2	Lat: 37 07′31.7″	Long: −122 03′37.5″
370733122033501	Loch Lomond Reservoir Thalweg A Range 27L2 near Ben Lomand	26	Lat: 37 07′33.9″	Long: −122 03′35.7″
370734122033601	Loch Lomond Reservoir Thalweg A Range 27L1 near Ben Lomand	26R1	Lat: 37 07′34.2″	Long: −122 03′36.0″
370734122033602	Loch Lomond Reservoir Thalweg A Range 26 near Ben Lomand	26R2	Lat: 37 07′34.5″	Long: −122 03′36.3″
370733122033502	Loch Lomond Reservoir Thalweg A Range 26L1 near Ben Lomand	26L1	Lat: 37 07′33.2″	Long: −122 03′35.2″
370732122033401	Loch Lomond Reservoir Thalweg A Range 25R2 near Ben Lomand	26L2	Lat: 37 07′32.5″	Long: −122 03′34.7″
370734122033101	Loch Lomond Reservoir Thalweg A Range 27 near Ben Lomand	27	Lat: 37 07′34.6″	Long: −122 03′31.7″
370735122033201	Loch Lomond Reservoir Thalweg A Range 26R1 near Ben Lomand	27R1	Lat: 37 07′35.2″	Long: −122 03′32.2″
370735122033202	Loch Lomond Reservoir Thalweg A Range 26R2 near Ben Lomand	27R2	Lat: 37 07′35.9″	Long: −122 03′32.7″
370733122033101	Loch Lomond Reservoir Thalweg A Range 27R1 near Ben Lomand	27L1	Lat: 37 07′33.9″	Long: −122 03′31.3″
370733122033001	Loch Lomond Reservoir Thalweg A Range 27R2 near Ben Lomand	27L2	Lat: 37 07′33.3″	Long: −122 03′30.9″
370736122033001	Loch Lomond Reservoir Thalweg A Range 'A' near Ben Lomand	28	Lat: 37 07′36.9″	Long: −122 03′30.8″
370737122033101	Loch Lomond Reservoir Thalweg A Range 28 near Ben Lomand	28R1	Lat: 37 07′37.1″	Long: −122 03′31.4″
370736122033002	Loch Lomond Reservoir Thalweg A Range 28L1 near Ben Lomand	28L1	Lat: 37 07′36.6″	Long: −122 03′30.1″
370737122032701	Loch Lomond Reservoir Thalweg A Range 29 near Ben Lomand	29	Lat: 37 07′37.5″	Long: −122 03′27.0″
370739122032501	Loch Lomond Reservoir Thalweg A Range 28R1 near Ben Lomand	30	Lat: 37 07′39.0″	Long: −122 03′25.0″
370740122032201	Loch Lomond Reservoir Thalweg A Range 30 near Ben Lomand	32	Lat: 37 07′40.7″	Long: −122 03′22.4″
370735122034201	Loch Lomond Reservoir Thalweg A Range 32 near Ben Lomand	A	Lat: 37 07′35.3″	Long: −122 03′42.7″

Analysis of Data Processing Methods

This section documents the data processing methods used to determine the most accurate estimate of storage capacity. These data processing methods were assigned on the basis of a combination of the individual data-point quality, the spatial distribution of the data, and the methods used to model the data. Finally, the change in water storage capacity since 1998 and the areas where sedimentation has occurred are documented.

Data Processing Analysis for Estimation of Storage Capacity

To determine the accuracy of each of the calculated capacity estimates, each of the methods and the data processing used for the calculations in the previous investigations were examined. The following explains each of the previous survey and current survey data processing methods, the accuracy of each method and, where applicable, alternate scenarios for data processing to estimate capacity with greater accuracy. As described previously in the Storage Capacity section, to accurately measure the storage capacity, all available data are required to most accurately represent the reservoir bed surface. To most accurately represent the reservoir bed surface, altitude data were integrated into a TIN model by using ArcInfo GIS software. For each of the new volumetric calculations, a surface model provided the necessary information to determine the capacity of the reservoir with the water level at its maximum (an altitude of 577.5 ft).

1960 Pre-Dam Construction

Because the original pre-dam construction potential reservoir capacity of 8,600 acre-ft was a rough estimate calculated from interpreted topographic contour lines interpreted from photogrammetry, the USGS wanted to improve this estimate of the potential storage capacity by using the topographic contour lines in addition to the dam structures to define the reservoir-bed surface. This method would allow data-processing techniques to be similar to those used in investigations completed in 1982 and later. Because the contour lines were not available from a digital source, a pre-dam construction, 10-ft-interval topographic contour map of the reservoir-bed surface survey was scanned and the rasterized map image was registered with known coordinates and converted into vectorized contours, which were then attributed with the altitude values from the map. The pre-1960 map does not include any of the dam structures, so estimated contours from the 1998 survey were used to place the estimated contours at the dam. These digital contour lines were used to estimate the original capacity with the approximate dam structure. This data-processing method

for determining capacity is similar to the method used in the 1998 investigation. Contour values and the thalweg location were used to guide the interpolation of altitudes in the area between contour lines. Additionally, where data were lacking, known spot altitudes were used. This new data-processing method indicated that the maximum storage capacity of the reservoir, at the spillway altitude of 577.5 ft, could have been about 8,770 acre-ft in 1960.

Because the average vertical accuracy of the photogrammetry is expected to be about 1.2 ft with a 95-percent confidence level (Federal Emergency Management Agency, 2006), the potential variances in storage capacity could be as much as 215 acre-ft, or within 2.5 percent of the total storage capacity. This resolution was determined by comparing reservoir-bed surfaces that represent the 1.2-ft potential error vertically. This estimated resolution does not take into consideration the potential error between contour intervals. The altitude of only about 8 percent of the reservoir-bed surface was defined with this method, leaving the altitude of the remaining area to be interpolated. Therefore, the revised resulting capacity at an estimated 95-percent confidence level is 8,770 ±440 acre-ft.

1971 Investigation

A new contour map of the reservoir surface was produced for this investigation even though the capacity estimates were based on the amount of sediment deposited on the lakebed. To better understand how the data-processing method of using contours to define a reservoir-bed surface could be compared with the previous investigation, the 1971 contour map was scanned and registered using the transect lines as a guide, and the capacity was calculated. The resulting estimate of 9,365 acre-ft, determined by using contour lines to define the reservoir surface, resulted in a capacity far greater than those calculated in any of the other investigations.

Measuring depth only along the transects and at important reservoir-bed surface features makes the determination of accuracy for this type of survey difficult. Accuracy of the vertical measurements is good but the spatial distribution of the data points could be so poor that it is not reasonable to make a storage capacity estimate. If it is assumed that a data point represents a 1-ft-radius area, the bathymetric survey data represent only about 0.5 percent of the total reservoir water-surface area.

As an alternative to calculating capacity from the revised contour map, it was calculated by subtracting the estimated volume of sedimentation from the estimate of reservoir capacity as of 1960. Brown (1973) reported that 46 acre-ft of sediment was estimated to have accumulated from 1960 to 1971. The resulting capacity, based on the original 1960 estimate of capacity of 8,600 acre-ft, is estimated with 91.5-percent confidence to be 8,554 ±795 acre-ft.

1982 Investigation

For the 1982 contour mapping, the survey transects were similarly placed and verified with the transformed contours as for the 1971 contour mapping, and new calculations of storage capacity were made, resulting in a calculated total capacity of 8,820 acre-ft. This estimate is erroneous, however, because data between the transects were insufficient (*fig. 6*)—the same limitation encountered in the 1971 survey. Features between transects were often missed, causing calculations of total capacity to be in error by as much as 1,000 acre-ft, as determined by modifying the contours in the areas with no data. When available data between transects were supplemented with 2009 altitude data, the revised resulting capacity, with an estimated confidence of 93.5 percent, is 8,820 ±870 acre-ft.

1998 Investigation

In 1998, altitudes at data points that could not be measured by boat using the GPS-enhanced depth echo sounder near the shoreline were defined by using a combination of spillway-water-surface altitude data points around the edge of the reservoir, spot altitudes measured at important features, and altitudes determined by differential leveling at data points along the thalweg of the inlet of Newell Creek. Therefore, a substantial area around the shoreline was poorly defined. Although the use of the GPS-enhanced depth echo sounder was a great improvement from previous surveys, portions of the reservoir bed were not surveyed. The published storage capacity determined by using data available at the time of the survey was 8,991 acre-ft. To improve the storage capacity calculations, the following scenarios were used to fill in missing data.

Scenario 1—Improve Shoreline with LiDAR Data

Data processing for the first scenario using the 1998 data set involved an attempt to fill in data where few or none were available along the shoreline. With the exception of the uppermost end of the reservoir, at the inlet of Newell Creek, the shoreline features showed little change from 1998 to 2009. With the assumption that the reservoir-bed surface changed negligibly along the shoreline, except at the inlet of Newell Creek, data from the 2009 investigation from an altitude of 575 ft to an altitude of 590 ft were used to add detail to the shoreline in the surface model. A recalculation using this information resulted in a calculated capacity of 8,735 acre-ft, which is 256 acre-ft less than the storage capacity initially calculated for 1998. This smaller capacity is likely a result of the initial 1998 water-surface perimeter being defined as larger than the true perimeter at the spillway altitude.

For this first scenario, with new shoreline data, the total capacity of Loch Lomond Reservoir in 1998 at a confidence level of 97 percent was 8,735 ±256 acre-ft rather than the published 8,991 acre-ft. Although the revised shoreline greatly improved the accuracy of the reservoir bed surface for 1998, some features that were missed during that survey appeared in the 2009 survey. Because some areas of the 1998 reservoir surface have not been defined and are subject to inaccurate interpolation, this scenario does not accurately represent the 1998 reservoir bed surface.

Scenario 2—Improve Shoreline with LiDAR Data and Missing Data with Sidescan Data

To determine the possible size of error in the original 1998 total storage capacity estimate, the missed features discussed above and illustrated by *figure 12* were included in a second scenario. The 2009 bathymetry and topography data sets were used to fill in the missing data. The 1998 data were preserved with a 50-ft-radius buffer, and altitudes in areas beyond this radius were supplemented with the 2009 data. An example of the data used in the data processing is shown in *figure 12*. This scenario, like the first, adds important shoreline data and features of the reservoir bed that may have been missed previously. Adding these missing data, however, results in a risk of estimating a storage capacity that is slightly smaller than the true capacity because the supplemental altitudes reflect sediment deposition as of 2009 and, therefore, a smaller storage capacity than actually existed in 1998. Scenario 2 results in a calculated capacity of 8,965 acre-ft—only 26 acre-ft less than the initial calculated capacity for 1998. This scenario reduces the potential error in the total storage capacity estimate and better represents the true capacity in 1998 and, therefore, is the most appropriate surface for comparison with the 2009 data. The resulting capacity of the reservoir in 1998 using scenario 2, with an estimated 97.5-percent confidence level, is 8,965 ±225 acre-ft.

2009 Investigation

Nearly all of the reservoir-bed surface altitudes measured throughout the reservoir with a state-of-the-art combination of topographic and bathymetric instruments could be incorporated into a reservoir bed surface model. The storage-capacity calculation for the 2009 investigation, based on this surface model with a 0.5-m (1.64-ft) horizontal distribution and vertical accuracy better than 0.1 m (0.33 ft), has a confidence level of 99 percent and is 8,646 ±85 acre-ft.

This reservoir bed surface model is sufficiently detailed to define most features that could substantially alter capacity calculations. And because the altitudes in the reservoir bed surface model have been spatially referenced to a common datum and coordinate, direct comparisons of the 2009 data with altitudes measured in future surveys can be accomplished easily.

To compare this new bed-surface model with that used in the 1998 investigation, several scenarios were examined to determine the scenario that would result in the most accurate comparison.

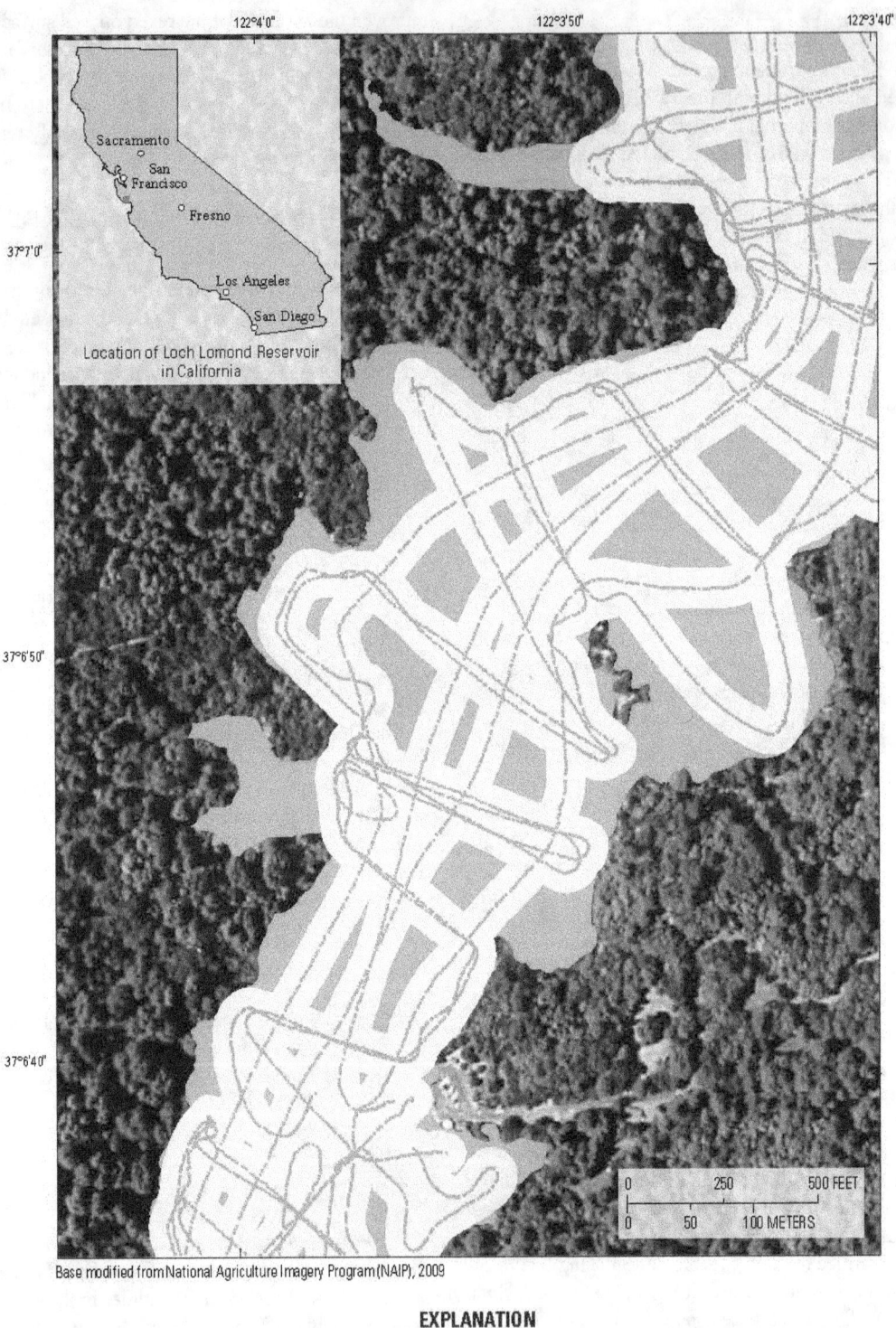

Base modified from National Agriculture Imagery Program (NAIP), 2009

EXPLANATION

Area of 2009 data collection used to supplement data from 1998 bathymetric survey
1998 bathymetric data-collection locations and 50-foot buffer area

Figure 12.   Example area in which 2009 bathymetric data were used to supplement data from the 1998 bathymetric survey of the reservoir-bed surface, Loch Lomond Reservoir, Santa Cruz County, California.

Scenario 1—Limit Data to 1998 Data Locations and Use Estimated Shoreline Data

For this scenario, the 2009 bathymetric survey was limited to the data locations of the 1998 bathymetric survey and to the limited shoreline data as described in the discussion of the 1998 investigation. To extract only the data from the 2009 survey that coincided with those in the 1998 survey, a buffer of 1-ft radius around the data points was used to select the subset of 2009 data points. These data were then combined with the poor shoreline-altitude data that were available for the 1998 investigation, resulting in a calculated storage capacity of 8,335 ±335 acre-ft at a 96-percent confidence level. The method used in this scenario is similar to the original method used to determine the 1998 storage capacity of 8,991 acre-ft. The resulting calculated capacity of the reservoir in 2009 by using scenario 1 is 630 acre-ft less than it was for 1998. This method does not include the accurate shoreline, however, and misses some of the reservoir-bed features, potentially causing error.

Scenario 2—Limit Data to 1998 Data Locations and Use Shoreline LiDAR Data

The second scenario limits the 2009 data to the locations of the 1998 bathymetry data, as previously described, but adds the detailed 2009 LiDAR shoreline data above an altitude of 578 ft. The result of using this scenario compares most closely with 1998 scenario 1 storage capacity of 8,735 acre-ft. This scenario for the 2009 data results in a calculated storage capacity, at a 97-percent confidence level, of 8,272 ±252 acre-ft, a reduction in storage capacity since 1998 of about 460 acre-ft. This scenario utilizes all the data available above the shoreline for both the 1998 and 2009 investigative periods, with assumptions of no substantial changes along the shore and altitude data that coincide throughout the reservoir bed, but replaces neither the erroneous data at the shoreline nor features that were missed in the data collection for the 1998 investigation. The reservoir-bed surface is not well represented in this scenario.

Scenario 3—Use All Bathymetry Data Available and Replace Shoreline Data with 1998 Estimated Data

Scenario 3 uses the well-defined bathymetry from the 2009 survey but with a poorly defined 1998 survey shoreline data. For this scenario, the bathymetry data up to an altitude of 575 ft were modeled with the limited shoreline data used in the 1998 investigation. This scenario produces a calculated storage capacity, at a 97-percent confidence level, of 8,675 ±265 acre-ft for 2009, which is much closer to the initial water-storage capacity of 8,646 acre-ft determined by using only 2009 altitude data. The results of this scenario are an indication that using poorly defined shoreline data can change the total storage capacity result by about 30 acre-ft.

The data processing method used in 2009 scenario 3 is most closely compared with the data processing method used in 1998 scenario 2. As previously described, the second scenario for 1998 used 2009 altitude data to fill in areas of missing data in the wetted reservoir bed and data up to the altitude of the top of the spillway; comparison with 2009 scenario 3 results in a water-storage capacity change that is smaller than the actual capacity change because of an under-estimate of capacity in areas of sedimentation in the 1998 investigation. This comparison indicates that there has been at least a 290-acre-ft reduction in storage capacity, owing to sedimentation, since the 1998 investigation.

All of these scenarios were evaluated and an approximation of the confidence level for each storage capacity was assigned based on a combination of the individual data-point quality, the spatial distribution of the data, and the methods used to model the data (*table 3; fig. 13*).

Capacity Change Since 1998

Although each of the data sets and data-processing scenarios has been examined, it is difficult to directly compare the total storage capacity values to determine the amount of sediment that has been deposited since 1998. However, because the data-processing method used in 1998 scenario 2 closely matches the data-processing method used in 2009 scenario 3 with the same confidence level, it can be safely estimated that a minimum of 290 acre-ft has been deposited. However, using the 1998 scenario 2 data-processing method to estimate a more accurate total storage capacity of 8,965 acre-ft, than that of the original estimate of 8,991 acre-ft, the storage capacity of Loch Lomond Reservoir decreased by 319 acre-ft to reach the March 2009 capacity of 8,646 acre-ft.

Sediment Data Analysis

As noted in the 1971 investigation report (Brown, 1973), sedimentation was calculated to be about 1,100 tons/yr/mi^2, which is comparable to annual sediment yields of 460 to 1,030 tons/yr/mi^2 reported for rural streams in nearby Santa Clara County. From November 1960 through November 1971, sediment deposition in Loch Lomond Reservoir caused a loss of storage capacity of at least 46 acre-ft (Brown, 1973). This loss of total storage capacity is considered a minimum, as indicated by the previously reported findings of the 1971 investigation. Also mentioned was that a layer of sediment 0.5-ft thick deposited over 180 acres of reservoir bottom would constitute 90 acre-ft of sediment, but to confirm this amount of sediment, the reservoir bed would need to be measured with very detailed surveying and extensive core sampling.

Table 3. Reservoir storage capacity and revised storage capacity estimates for Loch Lomond Reservoir, Santa Cruz County, California.

[DEM, digital elevation model; GIS, geographic information system; GPS, geospatial positioning system; IHO, International Hydrographic Office; LIDAR, light detection and ranging. **Abbreviations:** <, less than; –, not calculated]

Investigation year	Data collection method and data processing notes	Published storage capacity, in acre-feet	Confidence level of storage capacity, in percent [1]	Estimated sedimentation, in acre-feet	Storage capacity scenario calculations, in acre-feet
1960	Aerial photogrammetry prior to dam construction. Data processing: methods for volume calculations were not found.	8,600	Unknown	–	–
1971	Survey along profiles established, and additional spot values. Data processing: sedimentation rates were calculated and a total accumulation of sediment was determined, but no storage capacity volumes were determined due to limited data from the prior survey.	–	–	46	–
1982	Survey profiles permanently established with fathometer with land values surveyed well above spillway altitude, along with additional land feature altitudes. Data processing: the total storage capacity was computed, using a method described by Eakin and Brown (1939), which uses the surface area at 10-foot contours to calculate the volume of each prismoid. The contour method was verified using the range method, described by Eakin and Brown (1939).	8,824	90	–	–
1998	Survey bathymetry with GPS enhanced electronic depth sounding throughout reservoir with concentration on 1982 profiles. Data processing: surface created in GIS software was defined by contour values as a softline, thalweg as a hardline, mass points from shoreline DEM contours, and spot values for island and shore features.	8,991	96	55	–
1998 (scenario 2)	Data processing: surface created in GIS software was defined by data points collected from the 1998 survey, shoreline data points collected in the 2009 LiDAR survey, and 2009 bathymetry data 50 feet from 1998 data where no data were available.	–	97.5	–	8,965
2009	Survey bathymetry with GPS enhanced multi-beam sonar and shoreline topography with boat-mounted LiDAR. Data processing: surface created in GIS software was defined by data points collected from the 2009 bathymetry survey and shoreline 2009 LiDAR survey.	8,646	99	–	–
2009 (scenario 1)	Data processing: surface created in GIS software was defined by contour values below a boat measureable altitude of 575 feet as a softline, thalweg as a hardline, mass points from shoreline DEM contours, and spot values for island and shore features.	–	96	630	8,335
2009 (scenario 2)	Data processing: surface created in GIS software was defined by 2009 data points coincident with the 1998 data points, thalweg as a hardline, and shoreline 2009 LiDAR survey above an altitude of 578 feet.	–	97	460	8,272
2009 (scenario 3)	Data processing: surface created in GIS software was defined by 2009 data points below a boat measureable altitude of 575 feet, mass points from shoreline DEM contours, and spot values for island and shore features.	–	97	290	8,675

Table 3. Reservoir storage capacity and revised storage capacity estimates for Loch Lomond Reservoir, Santa Cruz County, California.—Continued

[DEM, digital elevation model; GIS, geographic information system; GPS, geospatial positioning system; IHO, International Hydrographic Office; LIDAR, light detection and ranging. **Abbreviations:** <, less than; –, not calculated]

Investigation year	Vertical resolution, in feet [2]	Spatial horizontal distribution, in percent [3]	Approximate accuracy range for storage capacity, in acre-feet [4]	Notes
1960	1.2	8	–	This survey was intended to provide an approximation to the volume of water Loch Lomond Reservoir would be able to contain so that water managers could determine the delivery amounts and the percent allocations. This range of storage capacity has been determined by comparing reservoir bed surfaces that represent the 1.2-foot potential error vertically. This does not take into consideration the potential error between contour intervals. Only about 8 percent of the reservoir surface is defined with this method, leaving the remainder to be interpolated.
1971	<0.5	0.5	–	No storage capacity determined. But an estimated 95-percent trap efficiency was determined by the 1971 study, with a sediment yield of 46 acre-feet (Brown, 1973).
1982	<0.5	0.5	7,940–9,710	As reported by Fogelman (1985), the inaccuracy of the base maps and initial surveys prohibits comparison with the 1982 computations; therefore, an estimate of the volume of sedimentation was not attempted.
1998	<0.1	3.4	8,630–9,350	At the time, this was an ideal survey because it located the altitude measurements without extensive data processing and recovery of profile lines; however, it lacked detailed shoreline data and some features were missed between the data collection points.
1998 (scenario 2)	<0.1	52	8,740–9,190	This scenario utilizes some of the 2009 data along the shoreline that are not likely to have changed since 1998 as well as altitude values at important features observed in 2009 that may have been missed between the data collection points. Adding the 2009 reservoir features may produce a capacity that is smaller than the true capacity as a result of sedimentation differences.
2009	–	100	8,560–8,730	This scenario represents ideal mapping of the reservoir bed surface and a surface resolution of less than 2 feet.
2009 (scenario 1)	3 (Within IHO Special Order)	3.4	8,000–8,670	Scenario 1 data processing method closely matches the 1998 data processing method. With a comparison of data using the same methods, the capacity has decreased by 630 acre-feet; however, the lack of data results in a poor representation of the reservoir bed and shoreline surface.
2009 (scenario 2)	3	4	8,020–8,520	Scenario 2 uses the same limitation of bathymetry data points as the 1998 survey, but with the LiDAR survey of the shoreline, giving an accurate definition of the shoreline. Again, the reservoir bed surface is not well represented.
2009 (scenario 3)	3	95	8,410–8,940	Scenario 3 uses the well-defined bathymetry from the 2009 survey but with a poorly defined shoreline, as with the 1998 survey. Shoreline features are missed and overly generalized, giving a greater storage capacity in the upper 10 feet.

[1] Confidence level estimates are approximations based on the data quality, data processing method, resolution, and data distribution. Mathematical representations for confidence level could not be determined.

[2] Resolution of altitude data derived from one or more types of data collection methods.

[3] Spatial resolution of data points on the reservoir bed as mapped two-dimensionally on the water surface. Each data point location is assumed to be surrounded by a buffer zone with a radius of 1 foot.

[4] Accuracy range based on confidence level of calculated or scenario storage capacity.

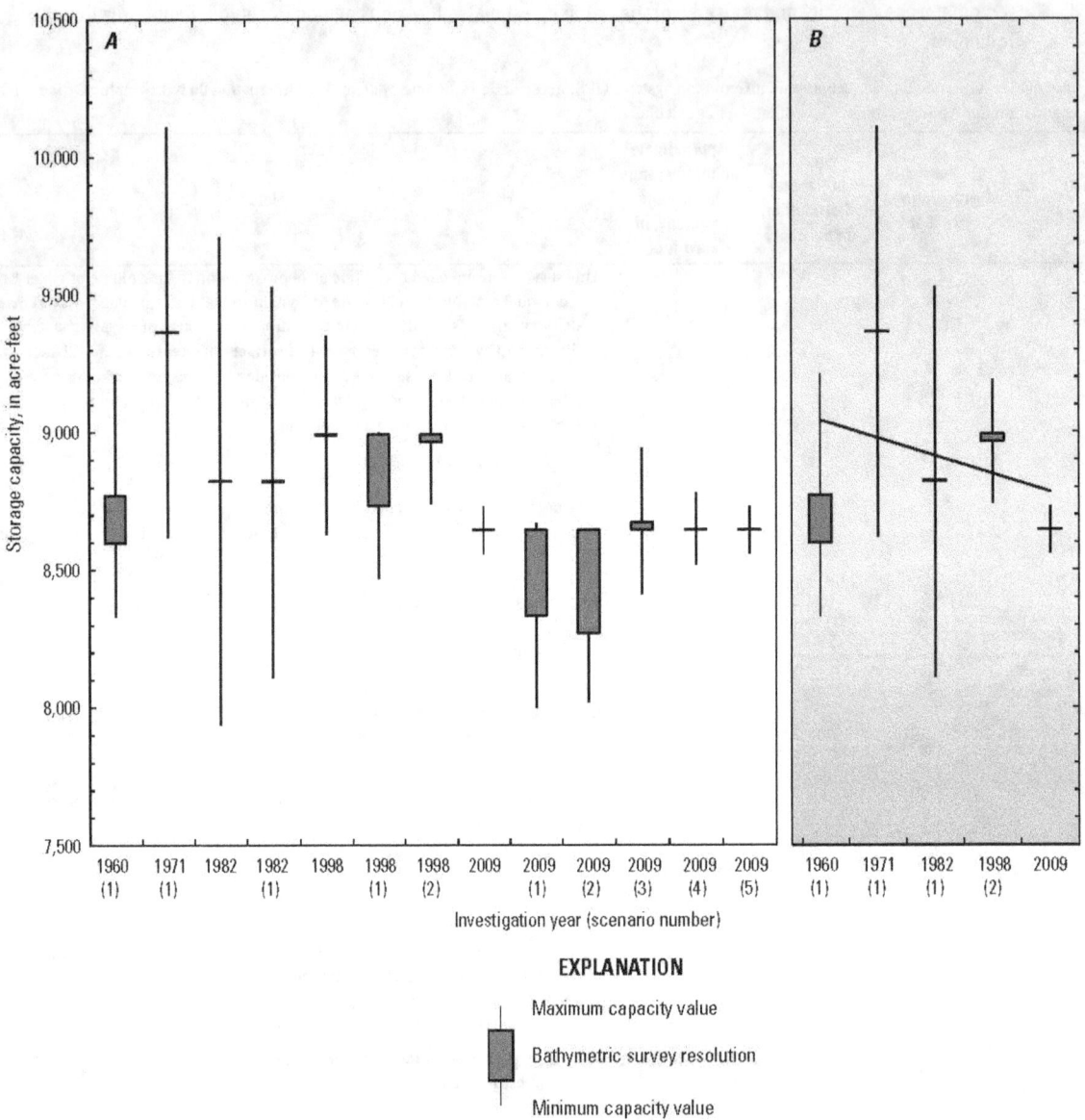

EXPLANATION

Maximum capacity value

Bathymetric survey resolution

Minimum capacity value

Figure 13.   Graphs showing (*A*) Minimum and maximum storage capacity, and total capacity resolution, estimated by evaluating the characteristics of the bathymetric surveys in each of the five investigations (1960, 1971, 1982, 1998, and 2009) and associated data-processing scenarios, and (*B*) estimated storage capacity ranges and resolutions for the indicated surveys and scenarios, and the storage capacity trend, Loch Lomond Reservoir, Santa Cruz County, California.

Se iment iel rom   to   co l not e eter mine   eca se o t e lac o com ara le ata Fogelman an o nson   n t e   in estigation it as concl e t at storage ca acit in t e   er reac o t e reser oir a een re ce   acre t since   o e er com arisons to earlier s r e s ere ro lematic eca se o i erences in meas rement tec ni es An eca se o limite a ail a le reso rces or meas ring reser oir e altit es at t e inlet o e ell ree to Loc Lomon eser oir in t e in estigation an a itional acre t o se iment ma a e een e osite in t e reser oir elo t es ill a altit e o   t t at a not incl e in t e calc lation

om arison o t e altit e ata rom t e c rrent in es tigation it t e limite altit e ata rom t e   in esti gation in icates t at se imentation as occ rre t ro g o t t e reser oir Se imentation t ro g o t t e reser oir as not re orte re io sl as a res lt o limite s r e ing tec ni es   ene met o o s r e ing as allo e or t e isco er o re io sl n no n se imentation an reser oir e eat res rom t e inlet o e ell ree all t e a to t e base of the dam. These new findings can lead to an improved etermination o t e rate o se imentation   allo ing t e total storage capacity of the reservoir to be quantified at a much finer resolution than in previous investigations. The

Table 4. Sediment size by size class.

[Blank cells indicate no value. No., number. **Abbreviation:** >, greater than]

Tyler sieve no.	U.S. standard sieve no. [1]	Class name	Metric units (millimeters)	Phi value [1] (θ)	English units (feet)
		Boulders	>256		>0.840
(²)	(²)	Large cobbles	256–128	−8	0.840–0.420
(²)	(²)	Small cobbles	128–64	−7	0.420–0.210
(²)	(²)	Very coarse gravel	64–32	−6	0.210–0.105
(²)	(²)	Coarse gravel	32–16	−5	0.105–0.0525
(²)	(²)	Medium gravel	16–8.0	−4	0.0525–0.0262
2.5	(²)	Fine gravel	8.0–4.0	−3	0.0262–0.0131
5	5	Very fine gravel	4.0–2.0	−2	0.0131–0.00656
9	10	Very coarse sand	2.0–1.0	−1	0.00656–0.00328
16	18	Coarse sand	1.0–0.50	0	0.00328–0.00164
32	35	Medium sand	0.50–0.25	+1	0.00164–0.000820
60	60	Fine sand	0.25–0.125	+2	0.000820–0.000410
115	120	Very fine sand	0.125–0.062	+3	0.000410–0.000205
250	230	Coarse silt	0.062–0.031	+4	0.000205–0.000103
		Medium silt	0.031–0.016	+5	0.000103–0.0000512
		Fine silt	0.016–0.008	+6	0.0000512–0.0000256
		Very fine silt	0.008–0.004	+7	0.0000256–0.0000128
		Coarse clay	0.004–0.0020	+8	0.0000131–0.00000656
		Medium clay	0.0020–0.0010	+9	0.00000656–0.00000328
		Fine clay	0.0010–0.0005	+10	0.00000328–0.00000164
		Very fine clay	0.0005–0.00024	+11	0.00000164–0.000000787

[1] For maximum size of the given class.

[2] Sieve openings are marked in inches and millimeters.

accuracy of these sedimentation rates is based on the resolution of the individual total storage capacity surveys being compared, or to the accuracy of the sediment study.

As expected in a reservoir's depositional environment (Colby, 1963), sediments deposited at the upstream end of the reservoir were coarser than those at the downstream end. As a result of their small size and slow fall rate in water (Guy, 1969), the silt- and clay-sized particles travel downstream in suspension until the fall rate of the particle overcomes the downstream-ward and upward convection velocity components of reservoir currents, and the particle settles to the reservoir bottom. Sediment particle size is generally discussed in terms of particle-size ranges (for example, sand, silt, clay; see *table 4*). Each descriptive size range is, in turn, composed of individual, physically (wet sieved) or theoretically (fall diameter) measured size classes (for example, coarse sand, medium sand, fine sand, very fine sand) (Guy, 1969).

The particle-size distribution results show that transect 2 (the downstream-most sampled location in 1982 and 2009) contained 77.7 percent clay and 5.5 percent sand in 2009, and 72 percent clay and 2 percent sand in 1982. The sediment sample from transect 32 (the upstream-most sampled location in 1982 and 2009) contained 2.8 percent clay and 73.6 percent sand in 2009, and 5 percent clay and 63 percent sand in 1982. In 2009, the sample from the transect 32 location also contained 13.2 percent fine gravel. In general, the particle-size distribution results indicate a slight coarsening of sediment over time throughout the length of the reservoir thalweg from 1982 to 2009, with the coarsening being most pronounced at the upstream end of the reservoir. A summary of results for samples collected at transect locations 2 and 32 during each of the sampling investigations is shown in *table 5*. Particle-size distribution data for all 36 samples collected during the 2009 study and for samples collected in previous investigations are presented in *appendix B* of this report.

Table 5. Particle-size distribution of bed sediment samples collected at transect locations 2 and 32, representing the downstream- and upstream-most sample locations, respectively, in 1971, 1982, 1998, and 2009, Loch Lomond Reservoir, Santa Cruz County, California.

[Particle-size values in percent; values may not sum to 100 due to rounding. **Abbreviations**: mm, millimeters; <, less than; >, greater than; –, no value]

Sample year	Clay (<0.004 mm)	Silt (0.004–0.062 mm)	Sand (0.062–2.00 mm)	Fine gravel (>2.00 mm)
		Transect 2		
1971	61	39	0	–
1982	72	26	2	–
1998	76	21	3	–
2009	78	17	6	–
		Transect 32		
1971	34	63	3	–
1982	5	32	63	–
1998	17	56	27	–
2009	3	10	74	13

Particle-size distributions of sediment samples collected along the reservoir thalweg (*fig. 14*) show an increase in the percentage of clay- and silt-sized particles with distance downstream. Heavier and coarser sands and gravels are deposited closer to the inlet of Newell Creek. The sample collected closest to the dam, at 62 ft upstream (transect 2, *fig. 14*) contained a smaller percentage of fine particles than the next closest sample, likely as a result of currents near the dam that maintain the fines in suspension and carry them to the outflow. At two locations in the lower portion of the reservoir (transects 6 and 10), sample-analysis results showed large amounts of sand, whereas the transect 2 sample contained a small amount. This discrepancy may be the result of unrepresentative sampling, but also may be a result of high-velocity storm flows that historically can extend far downstream. Flows during 1983, 1986, and 1998, as well as several other years, were relatively high (*fig. 4*). The upstream-most samples reflect the shallow, lower velocity flows that deposit nearly all sands.

Particle-size distributions of samples collected at the 36 sediment sites (*fig. 15*) show, as expected, that more sand is deposited in the uppermost reach of the reservoir than farther downstream. Finer silts and especially clays are deposited farther downstream (*fig. 15*). In the uppermost reach, the larger and heavier particles are deposited farther downstream as the velocity increases; as the velocity decreases, finer particles such as silts likely are being deposited over these sand particles. In addition to illustrating the fining of sediment with distance downstream through the length of the reservoir, *figure 15* also shows a representation of the path of dominant flow velocity in the upstream portion. For example, in the

uppermost reach near the northern shore, where velocities are sufficient to maintain the fines in suspension, deposits are dominated by sand. Fines are more prevalent along the opposite shore. Samples in transects 26 and 27 are mostly sand, deposited as the water curves around the slight bend, resulting in a shallow sand bar. Proceeding downstream, samples 10 and 6 are from locations where the reservoir narrows, likely resulting in higher velocities that maintain fines in suspension. The sample collected farthest downstream, near the dam, is dominated by silts, as the clays likely move out of the reservoir in suspension.

Reservoir Transects

Analysis of the cross section along each of the transects (*appendix C*; transect locations shown in *fig. 5*) provides information about the changes that have occurred throughout the reservoir. Depth of sediment in transects 2 to 7 had previously changed substantially, primarily as a result of landsliding within the wetted surface of the reservoir bed. This same reach of the reservoir showed changes between 1960 and 1998; from 1998 to 2009, the changes in depth clearly are a result of sedimentation rather than additional landsliding as shown by the uniform deposition across each of the transects. This range includes transects where depth has changed from 2 to 6 ft since 1998, with a change in cross-sectional area of 3.5 to 8.0 percent.

Transects 2, 2.5 (for 1960 data), and 17 (for 1982 data) (*appendix C*) are examples of transects that may not have been accurately located as a result of the survey techniques available at that time. Comparison of these cross-sectional profiles of altitude data in relation to those profiles from the GPS-located altitude data of 1998 and 2009 indicate that the 1960 and 1982 transects likely are spatially placed incorrectly as evident of significant features unlikely to have moved since 1982. These poorly placed transects could affect the calculations of total capacity in those investigations.

Transect 8 follows a narrow section of the reservoir where cross-sectional area increases about 60 percent. This wider section will cause flow velocity to decrease and additional sediments to settle. Particle sizes in samples 4 to 6 (*fig. 15*) support this assumption. This transect also is an example of a poorly defined shoreline in the 1982 investigation because the altitude measurements made along the right bank during both the prior and the subsequent investigation were similar to one another, but different than the 1982 measurements.

Sedimentation in the middle range of the reservoir, from about transect 9 to transect 20, is minimal. Change in depth at the thalweg is only about 0 to 5 ft along transects 9 (1971 survey) through 17.5. Transects 18 to 20 show almost no sedimentation—an indication that flow through this range is primarily near the reservoir bottom, so that the suspended sediments move farther downstream in the reservoir. The heavier sediments already have been deposited upstream.

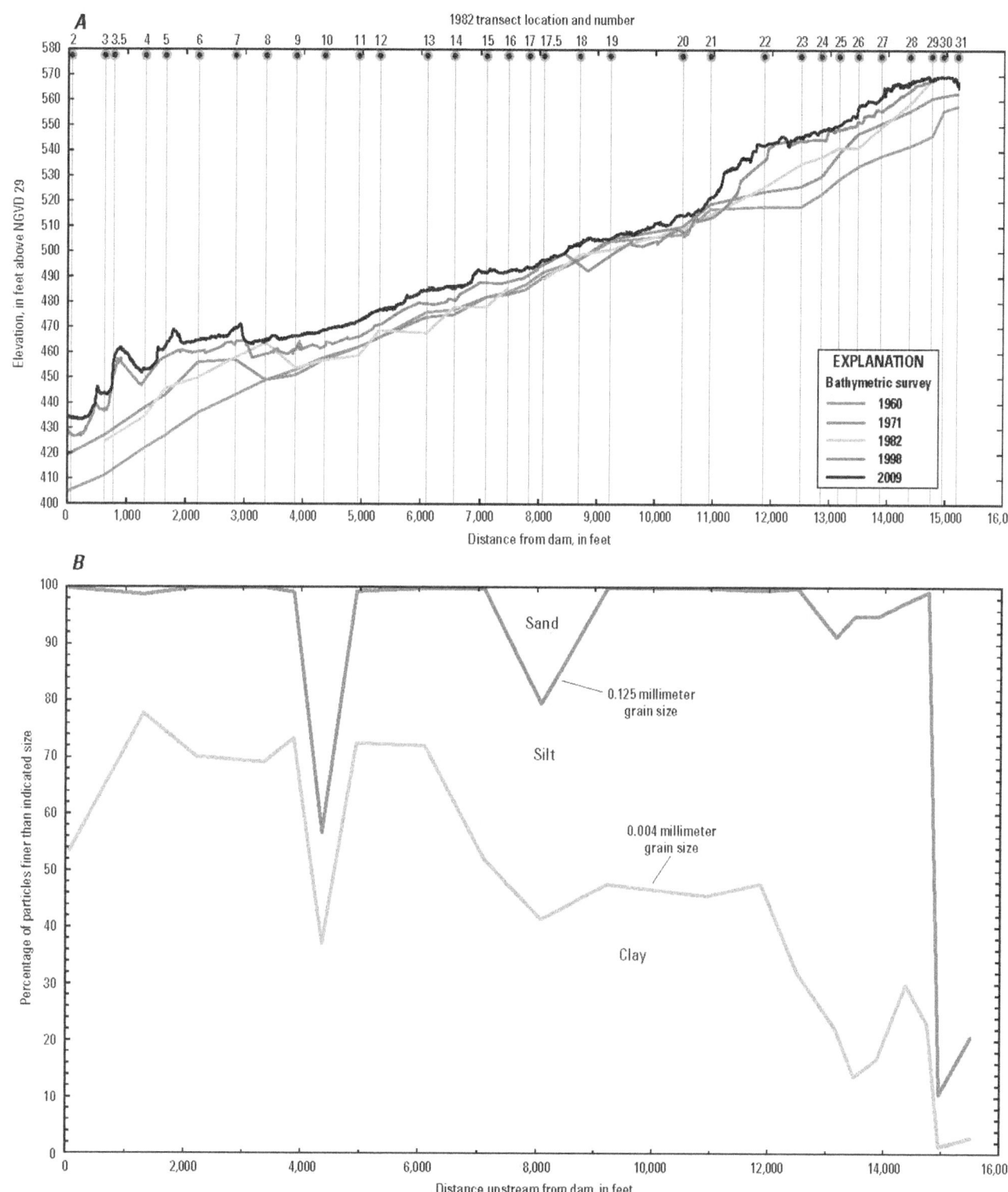

Figure 14. (*A*) Comparison of thalweg altitude for five bathymetric surveys (1960, 1971, 1982, 1998, and 2009), and (*B*) relation of particle-size distribution of bed-sediment samples collected May 19, 2009 as compared with distance from the dam along the thalweg, Loch Lomond Reservoir, Santa Cruz County, California.

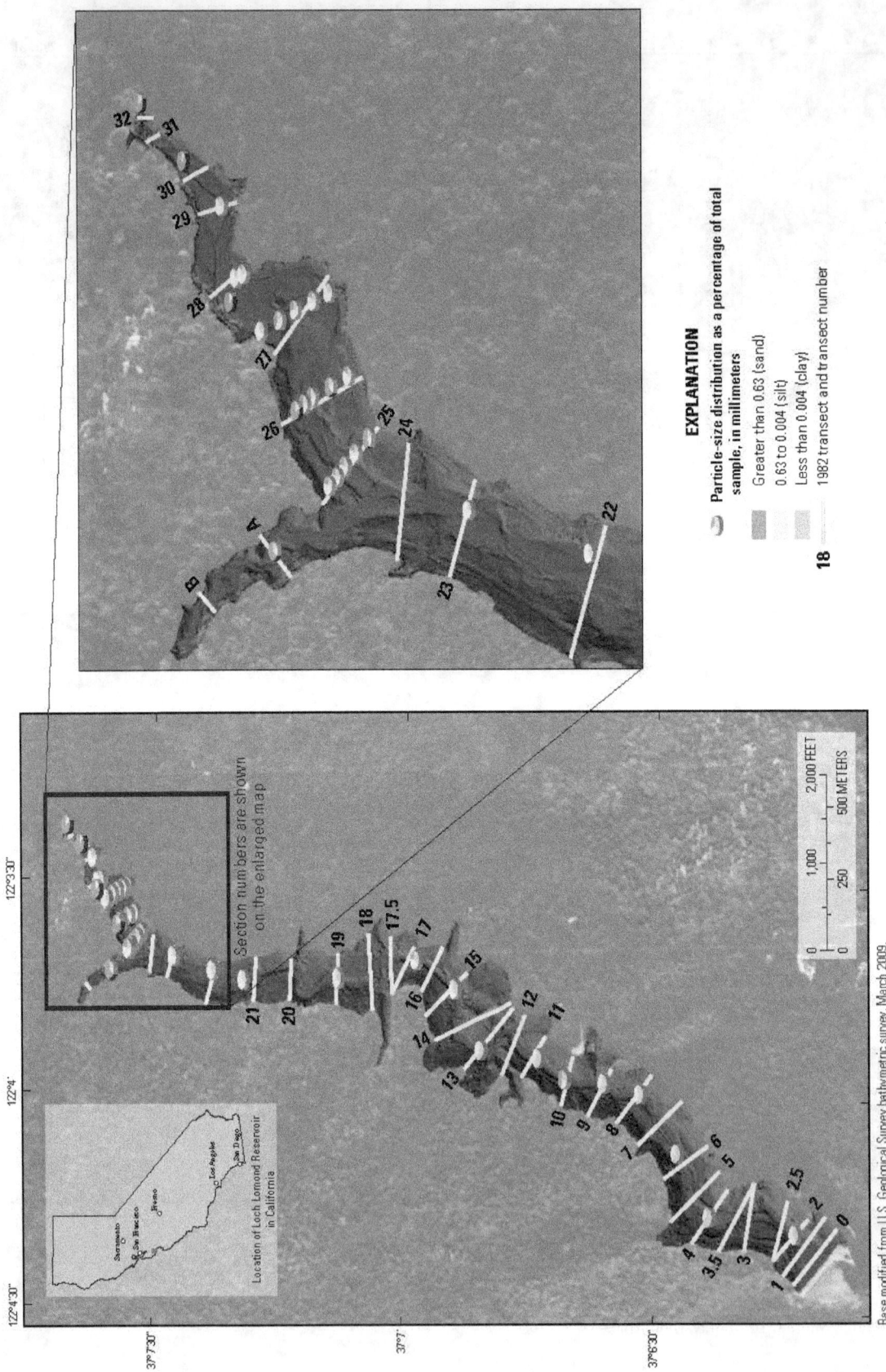

Base modified from U.S. Geological Survey bathymetric survey, March 2009,
and National Agriculture Imagery Program (NAIP), 2009

Figure 15.   Area distribution of particle sizes (sand, silt, and clay) as a percentage of individual bed-sediment samples collected May 19, 2009, in Loch Lomond Reservoir, Santa Cruz County, California.

Transects from 21 to the inlet of Newell Creek show sediment deposition, with the greatest change in channel shape in the uppermost portion of this range. Transects 21 through 30 show consistent sediment deposition from 1960 to about 1998. After 1998, sediment deposition increased only slightly in this reach because more sediment is carried downstream than at other transects. From about transect 29 upstream beyond transect 32, the reservoir is subject to the greatest amount of erosion during high inflows and of sediment deposition when the water surface is high. Consequently, the thalweg is likely to migrate from season to season.

At the upper end of the reservoir (above transect 21), sedimentation occurs as a result of the sediment discharge from Newell Creek and the small tributary, MacFarlane Creek, just downstream. Between 1960 and 2009, about 20 ft of sediment was deposited between transects 21 and 23, and about 30 ft of sediment was deposited between transects 23 and 28. During this period, the slope of the thalweg steepened from about 32.8 to 36.8 ft/mi in the reach above transect 21. Sedimentation in the remainder of the reservoir (between transects 2 and 21) decreased the slope of the original thalweg from 54.2 to 29.4 ft/mi (*fig. 14*).

Sedimentation and Erosion in Upstream Reach

Sedimentation continues to increase in the upstream reach of the reservoir along with erosion. This additional sedimentation is causing the area of deposition, representing a delta front, to extend about 500 ft farther downstream than at the time of the previous investigation (1998), during which the formation of the delta front was first measured. Sedimentation in the upstream reach is also causing the shoreline at the inlet of Newell Creek to widen in some areas. This erosion at the inlet is a result of the incision of new channels by substantial flows into the reservoir after smaller flows have deposited sediment in the thalweg and created the flat delta area. This process is evident in the changes to the reservoir-bed surface since 1998 (*fig. 16*). Comparing the 2009 reservoir bed surface above transect 25 with the second scenario revised reservoir bed surface from the 1998 investigation shows that about 15 acre-ft of erosion has occurred. Comparing these two reservoir-bed surfaces from the inlet of Newell Creek to about transect 22 indicates that almost 50 acre-ft of erosion has occurred.

Sedimentation throughout the Reservoir

This investigation indicates that some sedimentation has occurred throughout the reservoir. A delta front may be present near the dam, as is apparent from the 1998 and 2009 surveys and from the sand collected at the particle-size sample 2 site (*fig. 14*). The landslides that occurred near this region may

explain the increased sedimentation between transects 3 and 7. Generally, transects show a more consistent disbursement of sediments, and the full reservoir-bed comparisons show sedimentation throughout the reservoir. The 1998 and 2009 modeled reservoir-bed surface areas were compared and mapped in *figure 16* to visualize the major differences and to show areas of sedimentation or erosion. Because data for the revised 1998 reservoir-bed surface area are still sparse in some small areas, the investigation focused on changes of greater than 2.5 ft. The 1998 data-point locations were buffered by 10 ft to capture a sufficiently large area to compare with the 2009 data. *Figure 16* shows the wide disbursement of sediment and (or) erosion that was not previously mapped.

Comparisons of the 2009 transect profiles with the transects measured in the 1998 investigation indicate sedimentation of about 50 acre-ft. This difference was calculated by subtracting the difference in volume between the 1998, scenario 2, and the 2009 reservoir storage capacities using the average end-area method as described earlier. The transects for 2009 average about a 9.6-percent decrease in the cross-sectional area from the original 1960 reservoir bed described earlier. By applying this percent change to the data from the 1960 investigation, the resulting sedimentation calculation is more than 800 acre-ft. This value does not coincide with the 1971 estimate of sediment deposition from 1960 to 1971 of 46 acre-ft. It is not clear whether the 1960 storage capacity was almost 800 acre-ft greater than the 8,646 acre-ft found for 2009, because the 1960 survey was less accurate than the 2009 survey. A sedimentation amount of 319 acre-ft was found by comparing the previously described revised 1998 scenario 2 reservoir-bed altitudes with the 2009 reservoir-bed altitudes, whereas 50 acre-ft of sedimentation was confirmed by comparing previous cross-sectional findings with the change in the transect profiles. As previously discussed in analyses of storage-capacity measurements, calculation using only the cross-sectional areas does not provide the accuracy needed to make this type of comparison. Therefore, the more accurate calculated amount of sediment deposited into Loch Lomond Reservoir is about 320 acre-ft.

Additional issues related to reservoir sedimentation that may occur in the long term include (1) channel aggradation upstream from the reservoir, (2) channel degradation of the stream system downstream from the reservoir, (3) localized landslide activity along the reservoir banks, (4) elevated turbidity levels in the reservoir induced by the presence of suspended sediment during periods of high runoff, (5) water-quality issues related to the presence of organic matter and trace metals sorbed on sediment particles, and (6) water-treatment problems related to the removal of sediment from hydraulic machinery and water-distribution systems.

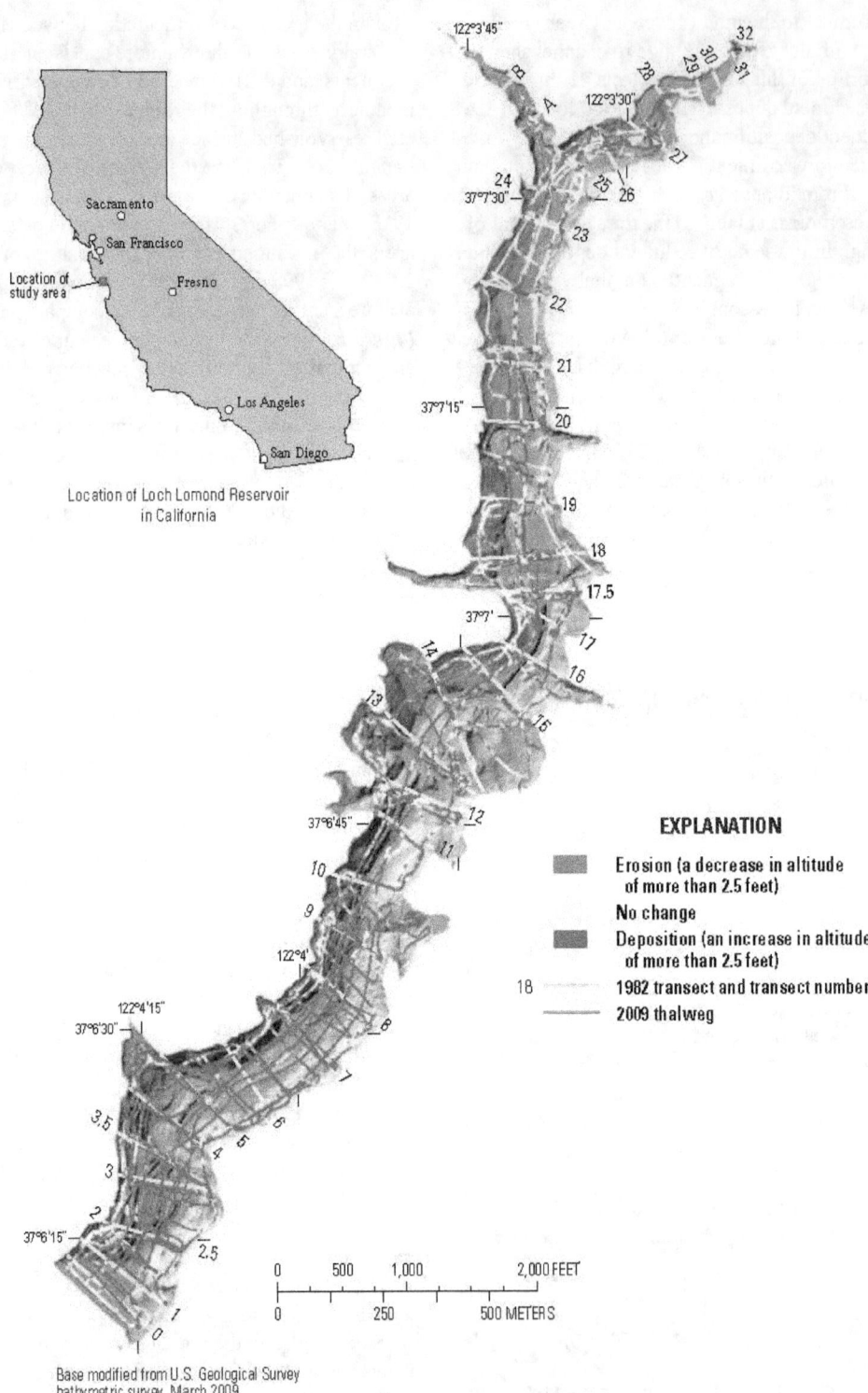

Figure 16.   Difference in altitude between 1998 and 2009 bathymetric surveys of Loch Lomond Reservoir, Santa Cruz County, California.

Summary

Loch Lomond Reservoir is an impoundment of Newell Creek in the Santa Cruz Mountains of California. It is owned by the City of Santa Cruz and is used as a source of water supply for the residents of the City. Since the reservoir was completed in 1960, park rangers and city water managers have routinely observed sedimentation at the inflow of Newell Creek to the reservoir. In March 2009, the U.S. Geological Survey (USGS), in cooperation with the City of Santa Cruz, conducted a bathymetric and topographic survey to determine the water-storage capacity, and the loss of capacity owing to sedimentation, of the reservoir. By using a new state-of-the-art method, which is a combination of bathymetric scanning using multibeam-sidescan sonar and topographic surveying using laser scanning (LiDAR) to measure altitude at data points on the reservoir bed surface, the USGS calculated a maximum storage capacity of the reservoir, as of March 2009, at the spillway altitude of 577.5 ft of 8,646 ±85 acre-ft with a confidence level of 99 percent.

This investigation was unique in that it offered the opportunity to repeat a series of measurements in the same location by using new and improved techniques in each decade since the reservoir was constructed. Although this approach highlights the uncertainties in the earlier methods, the repeated measurements allow for successive refinements in the collection and interpretation of altitude data in the reservoir environment. Many different methods, each using different equipment and different techniques for processing the altitude data, and therefore having a different associated accuracy, have been used over the history of Loch Lomond Reservoir to monitor storage capacity and rate of sedimentation. In an effort to accurately determine the storage capacity of the reservoir, the USGS, in cooperation with the City of Santa Cruz, examined each of these data-collection and data-processing methods to estimate the accuracy of the storage capacity calculated in each previous investigation, used new findings to correct those capacity estimates, and determined the most repeatable and cost-effective approach for the continued monitoring of the reservoir-bed surface.

Results of the current (March 2009) survey indicate that the maximum storage capacity of the reservoir at the spillway altitude of 577.5 ft, at a confidence level of 99 percent, was 8,646 ±85 acre-ft. This new, state-of-the-art method of altitude data collection and processing produced a high-resolution grid to determine a calculated total water storage capacity at the 99-percent confidence level. The state-of-the-art bathymetric and topographic method used for this survey accurately captures the features of the wetted reservoir-bed surface, as well as features along the shoreline that affect the storage capacity calculations.

The bathymetric and topographic surveys were performed in late March 2009, when the reservoir was at an optimum altitude of 577.5 ft that allowed boat access throughout the lake. Bathymetry of the reservoir bed was surveyed by using a boat-mounted SEA SwathPlusH 468-kHz interferometric bathymetric sidescan-sonar system, and data for the topographic survey were collected with a boat-mounted mobile laser scanning system (LiDAR).

Measuring sedimentation throughout the reservoir is difficult without a detailed survey of the reservoir bed and an understanding of the watershed history, including history of climate, fire, land-use change, earthquakes, forest logging, and road construction. Historic record of watershed activities provide a general idea of how much sediment could be expected, where detailed surveys provide accurate storage capacity values, and allow reservoir-bed surfaces measured in different years to be compared. Small features left unmeasured can substantially alter the total storage capacity estimate. Moreover, the reservoir-bed surface data for the different years need to be of equal quality and spatial distribution to provide an accurate determination of sedimentation by calculating the difference between the altitudes of the two surfaces. As in the previous investigations of the reservoir, the volume of sedimentation is considered equal to the decrease in water storage capacity. To determine the recent (1998–2009) change in storage capacity of Loch Lomond Reservoir, storage capacity volumes from the reservoir bed surface to the spillway altitude in those two years were compared. Comparison of the reservoir-bed surface defined by the March 2009 survey with a revised November 1998 reservoir-bed surface defined by using data from the November 1998 survey combined with new data indicates that sedimentation is occurring throughout the reservoir and totaled about 320 acre-ft from the 1998 to the 2009 investigation. This volume is about 3.5 percent of the total storage capacity estimated in the 1998 investigation.

Results of sampling and analysis of the reservoir-bed sediments show an increase in fine sediments farther downstream than in previous investigations, thereby supporting the conclusion that the storage capacity of the reservoir has changed owing to sediment deposition. Fine sediments are carried farther downstream as flow increases; when the water level is low, however, sediments eroded from the upper reach are carried farther downstream. Initially, sediment deposition is greater in the upper reach of the reservoir; as the water level falls, sediments in the upper reach are eroded and carried downstream, leaving the larger grained sediments upstream. Results of sediment sampling completed in May 2009 support the observation that erosion occurs in the upper reach, fine sediments are deposited throughout the remainder of the reservoir, and overall, the amount of sediment in the reservoir increases over time.

Analysis of this investigation indicates that the advanced method used in the 2009 survey accurately captures the features of the wetted reservoir surface as well as features along the shoreline that affect the storage capacity calculations, and because the bathymetric and topographic data are referenced to a datum, the results can be easily replicated or compared with future results. The techniques employed in this study to improve understanding of the quantitative effects of increased sedimentation rates may allow for a more effective assessment of changes in water-storage capacity in other, similar basins and reservoirs.

Future Bathymetric Surveys

The method used for the 2009 investigation has proven to be the most accurate method of documenting the reservoir-bed surface. This repeatable sidescan-sonar survey provides the necessary baseline data with which to compare the results of future surveys. Therefore, it would be advantageous to use the March 2009 bathymetric and topographic combined data set as the baseline for comparison with future bathymetric surveys.

It may not be necessary to repeat shoreline surveys as often as the bathymetric surveys, except in the upper portion of the reservoir where boat access is limited or when known shoreline movement has occurred. However, many factors may alter the shoreline altitudes measured during the LiDAR survey of the March 2009 investigation.

Because the continuously monitored water levels measured at staff gages provide information about the current storage capacity, verifying staff gage values will help to maximize the accuracy of the calculated storage-capacity values. These staff gages at the dam and boat ramp could be tested for accuracy by using conventional leveling methods and referencing to a benchmark. Additionally, installing a GPS-accessible benchmark near the boat ramp parking lot would provide a reference for the staff gage as well as a supplemental reference for future surveys.

Sedimentation Monitoring

Measurement of suspended sediment entering and exiting the reservoir helps support the findings of sediment deposition determined from reservoir bed comparisons. Continuous measurement of the sediment entering and exiting the reservoir would provide the data needed to calculate sediment budgets and predict future sedimentation. Installing monitoring stations (1) at the inflow of Newell Creek to Loch Lomond Reservoir, (2) at the inflow of MacFarlane Creek to Loch Lomond Reservoir, and (3) at the outflow of Newell Creek Dam would supplement available information about the characteristics of the sediment entering the reservoir, the amount of sediment being trapped by the reservoir, and the amount of sediment in the outflow from the reservoir, and therefore would help to refine current knowledge of the trap efficiency of the reservoir and help water-supply managers plan for sediment deposition monitoring.

Monitoring sediment discharge could include sampling and computation of both suspended sediment and bedload discharges. Results of lab analyses of these samples to determine the particle sizes being transported from the tributaries could then be compared with analyses of reservoir-bed sediment samples collected seasonally from the known delta formation areas within the reservoir to determine the potential rate of sedimentation. Bedload sampling most likely would not be needed as part of sediment monitoring downstream from the reservoir, as the larger particles presumably would be trapped in the reservoir. Various scenarios, each with different sampling schemes, computation methods, and costs could be employed. The scenarios could range from simply collecting a few sets of samples during storm events to develop general transport curves and compute gross estimates of annual sediment discharge, to a program of intensive sampling and computation of daily and (or) storm-event sediment discharge.

Installation of Telemetered Gages

The installation of a telemetered reservoir gage would provide water managers with near-real-time reservoir water levels and, therefore, accurate estimates of reservoir contents. Access to real-time reservoir data would allow city water managers to remotely manage the reservoir water supply in a more timely manner and in coordination with the other water resources in the city's water-supply system.

Another possibility would be to install telemetry on the streamgages associated with sediment monitoring on Newell and MacFarlane Creeks. Near-real-time inflow data could be a valuable tool, especially during periods of large or sustained runoff, to monitor potential for flooding or to make diversions to avoid uncontrolled reservoir spills. The streamgages could be enhanced by installing a suite of telemetered water-quality sensors to monitor tributary input of constituents that potentially could affect the quality of the drinking-water supply.

Installation of one or more precipitation gages at selected locations in the watershed would provide near-term benefits by alerting city water managers to high-intensity precipitation events, and increase knowledge of changes in the seasonal distribution of precipitation. Installation of precipitation gages coupled with sediment monitoring could also increase knowledge of precipitation effects on erosion and subsequent sediment deposition.

After review by the USGS, all telemetered data could be made available to the public on the USGS real-time Web page (*http://waterdata.usgs.gov/ca/nwis*).

Acknowledgments

The authors thank Rikk Kvitek, director of the California State University, Monterey Bay, Seafloor Mapping Lab, for providing the equipment used for the 2009 survey. Dr. Kvitek, his staff, and students aided in the operation of the equipment, developed the methodology for the survey, and processed the 2009 altitude data collected. Their surveying expertise, quality-assurance verification, data processing, and documentation substantially shortened the time needed to complete this investigation.

The authors thank the park staff at Loch Lomond Reservoir and the City of Santa Cruz for providing easy access to the reservoir during the survey and for providing their boat for sediment sampling.

The authors would also like to thank the USGS sediment-sampling staff, who altered their schedules to accommodate the sampling at ideal lake conditions; the reviewers who provided their knowledge and expertise to enhance the quality of this report; and the editorial staff that makes our science clear and understandable to all.

Selected References

Banks, W.S.L., and LaMotte, A.E., 1999, Sediment accumulation and water volume in Loch Raven Reservoir, Baltimore County, Maryland: U.S. Geological Survey Water-Resources Investigations Report 99-4240, 1 sheet. (Available at *http://md.water.usgs.gov/publications/wrir-99-4240/*)

Brabb, E.E., Graham, S.E., Wentworth, C., Knifong, D., Graymer, R., and Blissenbach, J., 1997, Geologic map of Santa Cruz County, California. A digital database: U.S. Geological Survey Open-File Report 97-489. Accessed October 23, 2008, at *http://pubs.usgs.gov/of/1997/of97-489/*

Brown, W.M., III, 1973, Erosion processes, fluvial sediment transport, and reservoir sedimentation in a part of the Newell and Zayante Creek Basins, Santa Cruz County, California: U.S. Geological Survey Open-File Report 73-35, 31 p. (Available at *http://pubs.er.usgs.gov/publication/ofr7335*)

Brown, W.M., III, and Knott, James, 1972, Topographic survey of Loch Lomond Reservoir: U.S. Geological Survey (with The Spink Corporation, Sacramento, California), 1 map sheet, scale 1:2400.

California Data Exchange Center (CDEC), 2009, California Data Exchange Center, Department of Water Resources, Precipitation. Accessed September 28, 2009, at *http://cdec.water.ca.gov/snow_rain.html*

California State University, Monterey Bay, Seafloor Mapping Lab, 2008, California seafloor mapping project. Accessed December 19, 2008, at *http://seafloor.csumb.edu/descriptions/laserdescrip.html*

Childs, J.R. , Snyder, N.P., and Hampton, M.A., 2003, Bathymetric and geophysical surveys of Englebright Lake, Yuba–Nevada Counties, California: U.S. Geological Survey Open-File Report 03–383, 20 p.

City of Santa Cruz, 1996, Urban water management plan: Santa Cruz, California, Santa Cruz Water Department, chap. 2, 9 p.

Colby, B.R., 1963, Fluvial sediments—A summary of source, transportation, deposition, and measurement of sediment discharge: U.S. Geological Survey Bulletin 1181-A, 47 p.

Cummings, J.C., Touring, R.M., and Brabb, E.E., 1962, Geology of the northern Santa Cruz Mountains, California: California Division of Mines and Geology Bulletin 181, p. 179–220.

Eakin, H.M., 1936 (revised by Brown, C.B., 1939), Silting of reservoirs: Washington, D.C., U.S. Department of Agriculture, Agricultural Technical Bulletin 524, 168 p.

Environmental Systems Research Institute (ESRI), 2008, ArcGIS, ver. 9.3: Redlands, California, ESRI. Accessed January 20, 2011, at *http://www.esri.com/software/arcgis/index.html*

Environmental Systems Research Institute (ESRI), 2009, ArcGIS online: Redlands, California, ESRI. Accessed January 20, 2011, at *http://resources.esri.com/arcgisonlineservices/*

Federal Emergency Management Agency (FEMA), 2006, National Flood Insurance Program (NFIP), Evaluation of alternatives in obtaining structural elevation data: U.S Department of Homeland Security, app. G, p. 287-288. (Available at : *http://www.fema.gov/library/file;jsessionid=BF995409A99D07B0784C6F508DC51F1E.WorkerLibrary?type=publishedFile&file=elevations_appg.pdf&fileid=c5046390-11ef-11e0-a151-001cc4568fb6*

Fogelman, R.P., and Johnson, K.L., 1985, Capacity and sedimentation of Loch Lomond Reservoir, Santa Cruz County, California: U.S. Geological Survey Open-File Report 85-485, 24 p. (Available at *http://pubs.er.usgs.gov/publication/ofr85485*)

Guy, H.P., 1969, Laboratory theory and methods for sediment analysis: U.S. Geological Survey Techniques of Water-Resources Investigations, book 5, chap. C1, 58 p. (Available at *http://pubs.usgs.gov/twri/twri5c1/*)

Guy, H.P., 1970, Fluvial sediment concepts: U.S. Geological Survey Techniques of Water-Resources Investigations, book 3, chap. C1, 55 p. (Available at *http://pubs.usgs.gov/twri/twri3-c1/*)

Heimann, D.C., 2001, Numerical simulation of streamflow distribution, sediment transport, and sediment deposition along Long Branch Creek in Northeast Missouri: U.S. Geological Survey Water-Resources Investigations Report 01–4269, 61 p. (Available at *http://mo.water.usgs.gov/ Reports/WRIR01-4269/index.htm*)

Heinemann, H.G., 1981, A new sediment trap efficiency curve for small reservoirs: Water Resources Bulletin, v. 17, no. 5, p. 825–830. (Available at *http://onlinelibrary.wiley.com/ doi/10.1111/j.1752-1688.1981.tb01304.x/pdf*)

Heinemann, H.G., and Dvorak, V.I., 1963, Improved volumetric survey and computation procedures for small reservoirs, in Proceedings of the Federal Interagency Sedimentation Conference, 1963: U.S. Department of Agriculture, Misc. Pub. No. 970, p. 845–856.

International Hydrographic Office, 1998, IHO standards for hydrographic surveys, 4th ed.: Monaco, International Hydrographic Bureau, Special Publication no. 44. Accessed October 3, 2009, at *http://www.imr.no/filarkiv/2006/08/S-44-eng.pdf/nb-no*

Langland, M.J., 2009, Bathymetry and sediment-storage capacity change in three reservoirs on the Lower Susquehanna River, 1996–2008: U.S. Geological Survey Scientific Investigations Report 2009–5110, 21 p. (Available at *http:// pubs.usgs.gov/sir/2009/5110/*)

Maidment, D.R., 2002, ArcHydro, 2002, GIS for Water Resources: Redlands, California, ESRI Press, 203 p.

McPherson, K.R., and Harmon, J.G., 2000, Storage capacity and sedimentation of Loch Lomond Reservoir, Santa Cruz, California, 1998: U.S. Geological Survey Water-Resources Investigations Report 00-4016, 16 p. (Available at *http:// ca.water.usgs.gov/archive/reports/wrir004016/*)

National Agriculture Imagery Program, 2009, Geospatial data gateway—2009 imagery, California: U.S. Department of Agriculture, Natural Resources Conservation Service. Accessed September 21, 2010, at *http://datagateway.nrcs. usda.gov/*

Smith, Douglas, 2010, Emerging technology for reservoir capacity studies: Water efficiency, July/August, p. 56–57. (Available at *http://www.waterefficiency.net/july-august-2010/emerging-technology-reservoir.aspx*)

Snyder, N.P., Rubin, D.M., Alpers, C.N., Childs, J.R., Curtis, J.A., Flint, L.E., and Wright, S.A., 2004, Estimating accumulation rates and physical properties of sediment behind a dam: Englebright Lake, Yuba River, northern California: Water Resources Research, v. 40, W11301, doi:10.1029/2004WR003279. Accessed October 20, 2009, at *http://www.agu.org/journals/ABS/2004/2004WR003279. shtml*

Swanson Hydrology & Geomorphology, 2001, Technical addendum to Zayante Area Sediment Source Study: Santa Cruz County, California, Swanson Hydrology & Geomorphology, 67 p. (Available at *http://sccounty01.co.santa-cruz. ca.us/eh/environmental_water_quality/pdfs/sediment_ study_appendices-final.pdf*

U.S. Geological Survey, 1999, Strategic directions for the Water Resources Division, 1998–2008: U.S. Geological Survey Open-File Report 99-249, 27 p. (Available at *http:// pubs.usgs.gov/of/1999/ofr99-249/html/exec.html*)

Vanoni, V.A., 2006, Sedimentation engineering: Reston, Virginia, American Society of Civil Engineers, manual 54, 415 p. (Available at *http://cedb.asce.org/cgi/WWWdisplay. cgi?0600308*)

Wilson, G.L., and Richards, J.M., 2006, Procedural documentation and accuracy assessment of bathymetric maps and area/capacity tables for small reservoirs: U.S. Geological Survey Scientific Investigations Report 2006–5208, 24 p., plus oversize figs. (Available at *http://pubs.usgs.gov/ sir/2006/5208/*)

Wilson, G.L., and Richards, J.M., 2008, Differences in reservoir bathymetry, area, and capacity between December 20–22, 2005, and June 16–19, 2008, for lower Taum Sauk Reservoir, Reynolds County, Missouri: U.S. Geological Survey Scientific Investigations Map 3061, 1 sheet. (Available at *http://pubs.usgs.gov/sim/3061/*)

Glossary

ArcInfo A full-featured geographic information system produced by ESRI, a software development and services company providing Geographic Information System (GIS) software and geodatabase management applications, that is the highest level of licensing (and therefore functionality) in the ArcGIS Desktop product line. ESRI's Web site can be found at *http://www.esri.com*

Average end area method A volume calculation that assumes volume between two consecutive cross sections is the average of their areas multiplied by the distance between them.

Bathymetric survey Survey of the measurable land surface below the water level. Many methods can be used for this type of survey. Typically the survey is completed by boat and depths are measured using either a weighted line, echo sounder, or other sonar device.

CSUMB SFML California State University, Monterey Bay, Seafloor Mapping Lab

Data set A combined group of data that may have been collected by different means but collectively have a specific purpose.

Digital Elevation Model (DEM) An elevation model created for use in computer software where bare-earth elevation values have regularly spaced intervals in latitude and longitude (x and y).

GIS Geographic information system.

Interferometric sonar The technique of superimposing (interfering) two or more waves to detect high- resolution differences between them.

LiDAR Light Detection and Ranging—A laser device that emits pulses, reflections of which are gathered by a telescope aligned with the laser. The return signal is used to determine distance and position of the reflecting material.

NWIS National Water Information System—A database maintained by the U.S. Geological Survey to store and view current and historical streamflow, groundwater-level, and water-quality data.

Sidescan sonar An acoustic imaging device used to produce wide-area, high-resolution backscatter images of the bed of a water body.

Staff gage Commonly a rugged iron gage with measurement markings and finished with a special porcelain enamel used for a quick visual indication of the surface level in reservoirs, rivers, streams, irrigation channels, weirs, and flumes.

Stage The height or altitude of the surface of a body of water above an arbitrary point or datum; one of a series of positions or altitudes.

Storage capacity As it relates to this report, the volume of water contained above the reservoir surface to a specific height, such as the altitude of the spillway.

Thalweg The deepest continuous channel along a valley or watercourse.

TIN Triangulated irregular network—A surface representation derived from irregularly spaced points and breakline features. Each sample point has an x, y coordinate and a z value or surface value.

Topographic survey Survey of the land surface, usually above ground. This type of survey can be completed using any of a large number of devices. A fast method of measuring large surface areas is to use LiDAR to record a large number of data locations with x and y coordinates coupled with an altitude.

Appendix A.

Draft Final Report, Loch Lomond Reservoir Capacity and Sedimentation Study, 2009: California State University Monterey Bay Seafloor Mapping Lab.

Draft Final Report
"Loch Lomond Reservoir Capacity and Sedimentation Study – 2009"
Cooperative Agreement G09AC00072
U. S. Geological Survey
California Water Science Center
and
California State University Monterey Bay Seafloor Mapping Lab

Full-Basin Topographic Survey of Loch Lomond Reservoir Using Vessel-mounted Topographic LIDAR and Interferometric Bathymetric Sidescan Sonar

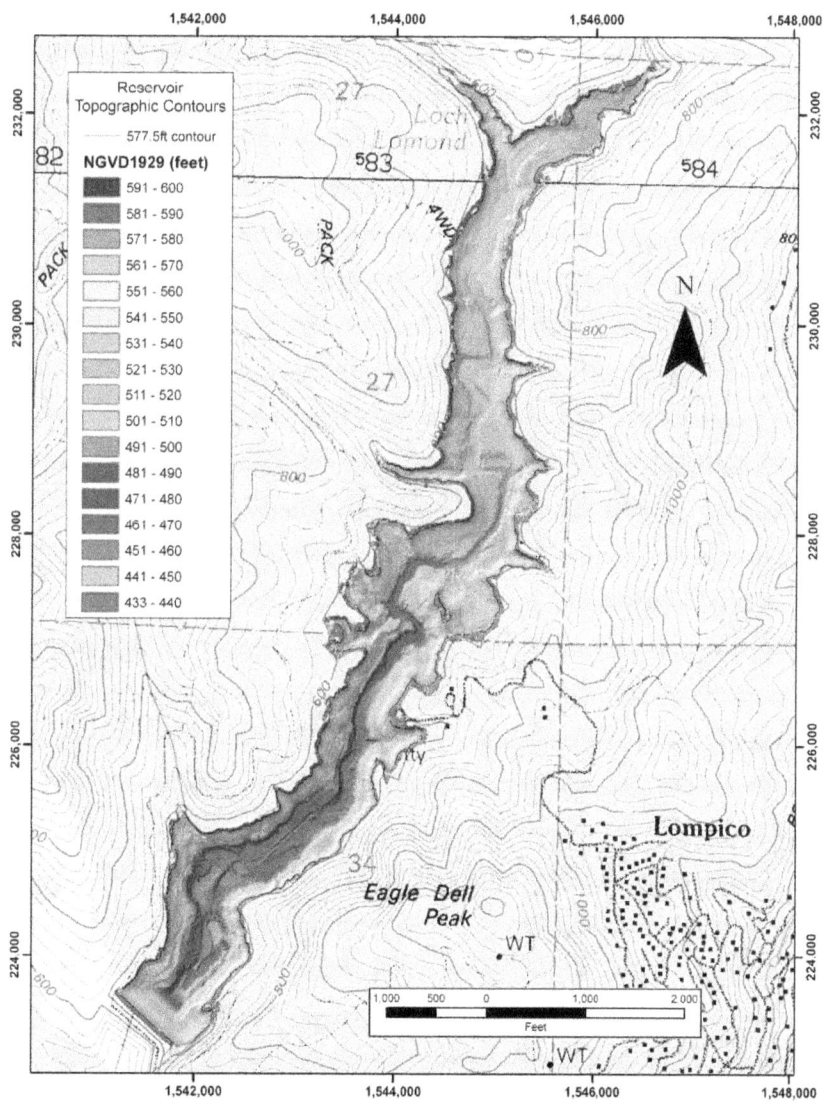

Survey dates: March 28-29, 2009
Prepare by: Rikk Kvitek, CSUMB Seafloor Mapping Lab
Preparation date: October 15, 2009

1 SUMMARY

In March of 2009, the CSUMB Seafloor Mapping Lab employed a novel combination of mobile vessel-mounted topographic LiDAR (terrestrial laser scanner) and interferometric bathymetric sidescan sonar for full-basin bare earth mapping of the Loch Lomond Reservoir in the Santa Cruz Mountains. The interferometric sidescan sonar system was able to map bathymetry from the reservoir floor up to the water surface, and the mobile laser scanner covered the exposed basin topography from the water surface up to the top of the reservoir's spillway retaining wall. Manual and automated cleaning and filtering were able to remove all water column debris and terrestrial vegetation yielding a comprehensive, high-density merged bathy/topo bare earth xyz point cloud containing > 130 million individual data points. These points were used to create final bare earth gridded xyz data sets of the entire 175 acre basin at 1.0 and 0.5 m resolution that met or exceeded IHO Special Order hydrographic survey standards.

TABLE OF CONTENTS

LIST OF FIGURES

LIST OF TABLES

2 PROJECT DESCRIPTION

2.1 Project Goal

The purpose of this project was to create an accurate, high-resolution digital elevation model (DEM) of the Loch Lomond Reservoir, an impoundment of Newell Creek in the Santa Cruz Mountains of California, based on newly acquired data in support of the 2009 "Loch Lomond Reservoir Capacity and Sedimentation Study" being conducted by the U.S. Geological Survey (USGS) for the city of Santa Cruz, California. Because the USGS and the California State University Monterey Bay (CSUMB) Seafloor Mapping Lab (SFML) share a mutual interested and desire to cooperate in understanding the natural and anthropogenic factors impacting the capacity and regulation of the California's reservoir-based water supplies, the USGS entered into Cooperative Agreement G09AC00072 with the University Corporation at Monterey Bay to: 1) conduct a comprehensive bathymetric and topographic survey of the Loch Lomond Reservoir, and 2) create a full-basin DEM from these survey data up to the level of the dam crest. Here we describe the specific objectives, methods, results and final products associated with and resulting from this cooperative agreement.

2.2 Background

Loch Lomond Reservoir, an impoundment of Newell Creek, is in the Santa Cruz Mountains, California and is owned by the city of Santa Cruz. The reservoir opened for public recreation in 1963, at which time it became a source of water supply for the city of Santa Cruz. Sedimentation has been observed by park rangers and city water managers at the inflow of Newell Creek for many rears. Water managers for the city of Santa Cruz periodically measure storage capacity to determine if any sedimentation has occurred, allowing water managers to take timely and appropriate actions. Sediment deposition has occurred in the lower reach of the reservoir because of landslides and in the upstream reach because of inflow from Newell Creek (Fogelman and Johnson, 1985). In 1982 and 1998, bathymetric surveys were completed to determine the storage capacity and the loss of capacity owing to sedimentation of Loch Lomond Reservoir. The volume of sedimentation in a reservoir is considered equal to the decrease in storage capacity. To determine sedimentation in Loch Lomond Reservoir, change in storage capacity was estimated for an upstream reach of the reservoir. Cross sections from the previous surveys were compared to determine the magnitude of sedimentation in the upstream reach of the reservoir. Results of the previous comparison, which were determined from changes in the cross-sectional areas, indicated that the capacity of the reservoir decreased by 55 acre-feet. To help water managers better regulate water supply, the new storage capacity of the reservoir in 2009 and changes in sedimentation need to be determined. The city of Santa Cruz has specifically asked the U.S. Geological Survey (USGS) to complete a new bathymetric and sediment survey of the reservoir to determine the change in capacity and the extent of sedimentation.

2.3 Work Plan

The goals of the cooperative agreement involve three key elements. 1) Data collection for the bathymetric and topographic surveys was to be completed in FY 2009 by boat when

the lake level was at an optimum level for boat access throughout the entire reservoir. The boat was to be provided either by the Loch Lomond Reservoir Park Ranger or by the SFML. 2) Ground surveys were to be used as necessary on recoverable bench marks established from previous surveys. Data for the bathymetric survey were to be collected using a multibeam sonar system, and data for the topographic survey were to be collected by a mobile laser scanning system (LiDAR). Expertise and equipment was to be provided by the SFML under the direction of Dr. Kvitek. Equipment was to be mounted on a boat operated by SFML staff for the survey. The bathymetric and topographic surveys were to take approximately 2 days on the water with additional time for survey setup. Results are to be delivered as XYZ locations in ASCII format, raw data in native data collection formats, and documentation in either ASCII or Word formats. Final datasets are to have any corrections applied and all erroneous data removed such as under water snags, schools of fish and other backscatter up to the high water mark. Raw datasets are to be provided to the U.S. Geological Survey and all data are considered public data and will be made available. A description of the data collection methods, corrections, manipulations, and quality assurance methods are to be documented and given to the U.S. Geological Survey to be used in a publication.

These goals are to be accomplished by the collaborative efforts of the USGS and the CSUMB Seafloor Mapping Lab scientists and students under the direction Dr. Rikk Kvitek. The studies offer a unique opportunity for both CSUMB and USGS to provide unbiased relevant information and data for use by Federal, State and local governments, municipal water managers and the general public. In conducting these studies, Dr. Kvitek and his research group worked towards the following specific goals:

- Dr. Kvitek will oversee all field, data processing and documenting activities.
- Assist in the planning and implementation of field studies.
- Provide expertise and research specialists to complete bathymetric and topographic surveys using equipment provided by the CSUMB SFML.
- Develop logistically practical methods for surveying the reservoir bed below the water level and above the water level up to the high water line.
- Topographic survey to be completed with a resolution of approx. 0.5 meter resolution or better and vertical accuracy with 10 cm.
- Bathymetric survey to be completed with a 0.5 meter resolution or better and vertical accuracy within 10 cm.
- Plan, develop implement and document quality assurance methods applied to topographic and bathymetric surveys.
- Abide by inspection and decontamination procedures as set by the Ranger of the Loch Lomond Recreation Area for boats and equipment used in the reservoir
- Apply any corrections and remove any erroneous data such as vegetation, schools of fish and backscatter to final datasets.
- Supply raw datasets. All data are considered public data and will be made available.
- Research group performing surveys will be experienced in boating and safety.
- Photos and videos taken of the survey process will be used to describe/illustrate methods and will be public data.

3 METHODS

3.1 Survey vessel and dates

The CSUMB Seafloor Mapping Lab provided the R/V MacGinitie as the platform for both the bathymetric and topographic surveys conducted at Loch Lomond March 28-30, 2009 (Figures 1 and 2).

R/V MacGinitie specifications:
Make and Model: SeaArk, Little Giant
Length: 32ft overall, with a 27ft hull length.
Draft: 1.5 ft.
Beam: 8.5ft
Fuel: 100 gals regular gasoline
Gross weight: 10,000 fully equipped.
Engines: twin Honda 130 hp, counter rotating, 4 stroke outboards that meet EPA emission standards for 2006. Top speed 34 knots. Survey speed 8-12 knots. Cruising speed 18-28 knots depending on conditions.
Electrical power: 30 amps 110VAC, 12VDC
Electronics: PC-based Nobeltec/Sitex navigation includes fully integrated GPS, digital charting, radar and autopilot.
Safety & radio equipment: EPIRB, life raft, flares, UHF radio, submersible GPS and UHF radios.

According to the park staff, the R/V MacGinitie is the largest vessel ever launched on the reservoir, and based on our experience towing the boat to the site, it would not be possible to get a larger boat up the access road to the reservoir due to the narrow hairpin turns along the way. The park provided a mooring dock and shore power for charging the inverter batteries overnight between the successive survey days.

Figure 1. R/V MacGinitie tied up at Loch Lomond floating docks in front of launch ramp and parking area.

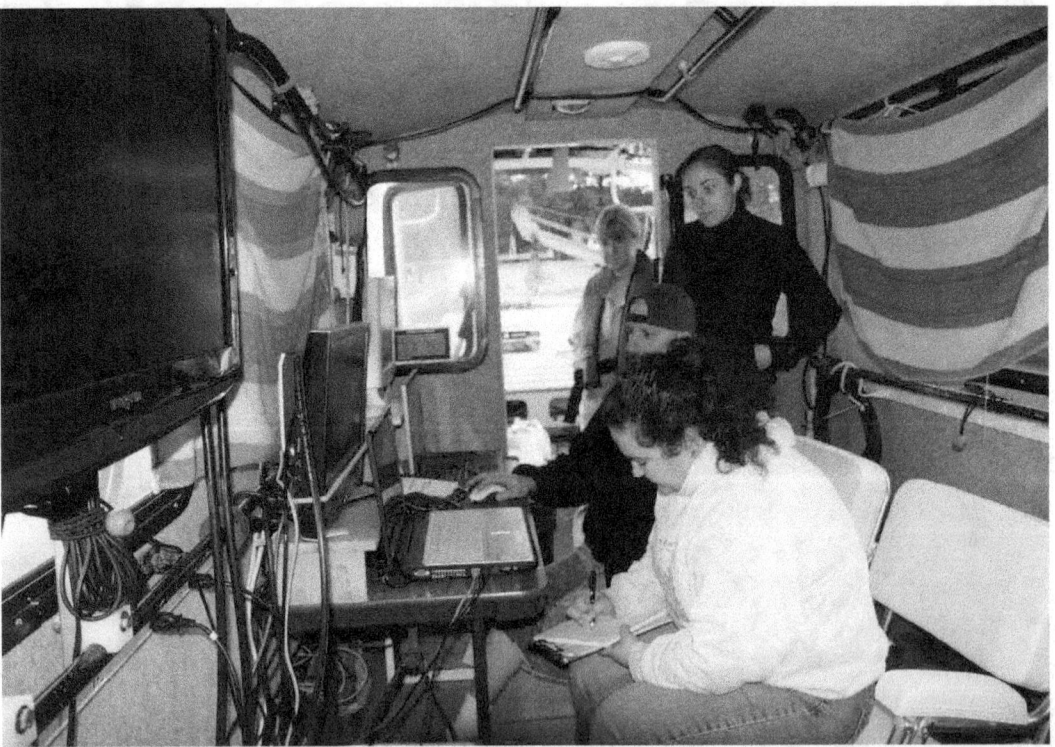

Figure 2. R/V MacGinitie interior configured for bathymetric data acquisition. Towels over windows are to minimize glare on the computer monitors used to control the sonar and survey navigation software. Personnel from back to front: Kelly McPherson (USGS project leader), and SFML hydrographers Katie Glitz, Pat Iampietro, Kate Thomas.

3.2 Positioning control

Three complementary methods of redundant position control were used during the surveys: 1) GPS reference station placed over a previously established benchmark on the dam crest was use as the master geodetic horizontal and vertical control for the project, 2) another previously established benchmark on the spillway retaining wall as used as a check on vertical control for the laser topographic results, and 3) the reservoir water level staff values were recorded at the beginning of the surveys as another accuracy check on vertical control.

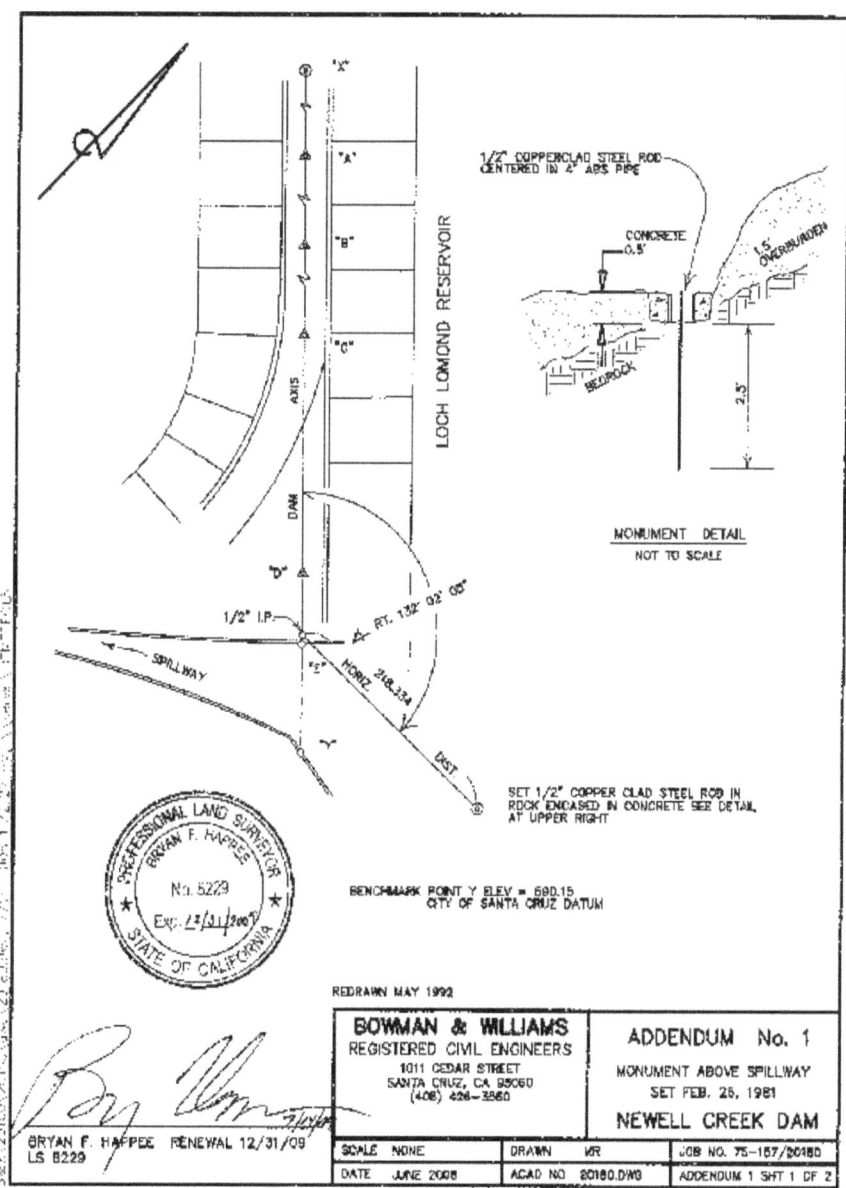

Figure 3. Field sheet describing and showing the location of the various benchmarks associated with the Loch Lomond dam site.

Figure 4. GoogleEarth view of the locations of the dam crest benchmark (yellow pin marker) used for GPS reference station and shown in Figure 7 below and water level staff gauge at the dam site (green dot) used to note reservoir water surface elevation during the survey (Figure 6).

Figure 5. GoogleEarth view of spillway benchmark (red dot) used for placement of reflective laser target shown in Figures 15-19 below.

Figure 6. Reservoir water level staff gauges at the dam site location shown in Figure 4. Lower staff shown in the photo is the one used to note water surface level for each day's survey.

3.2.1 GPS reference station

Geodetic survey control for both the laser and sonar surveys was based on the post-processed solution from a GPS reference station placed over an established benchmark atop the dam crest.

Figure 7. Trimble NetR5 GPS reference station L1/L2 antenna set up over bench mark on dam crest referenced in the field sheet shown in detail in Figure 11 below. Red arrows point to the top of the copper clad steel rod benchmark inside the recessed concrete collar and the reference location on the field sheet.

A Trimble Navigation NetR5 dual frequency reference station with a TRM55971.00 L1/L2 antenna (Figures 7 and 8) was setup to log data files at 1 Hz before the beginning of each day's survey and taken down at the end of the day. The dam benchmark site was accessed from the water using a small inflatable launch from the MacGinitie and rowed ashore to the dam (Figures 9 and 10). The antenna height was measured each day from the top of the benchmark to the bottom of the antenna notch after setting up the antenna tripod.

28 march 2009: swathplus bathy survey
Antenna height measured to bottom of notch: 1.266m

29 march 2009: laser scanner topo survey
Antenna height measured to bottom of notch: 1.420m

Back in the lab, RINEX files were created from original .T01 base files using Trimble Convert2rinex application. The Antenna height listed in Rinex file is the ARP height calculated by Convert2rinex. This Rinex ARP value matched the ARP height calculated independently using trig from the:

slant height measured in field
antenna radius taken from spec picture on antenna bottom & NGS image file
distance between bottom of antenna and bottom of notch based on specs and NGS image.

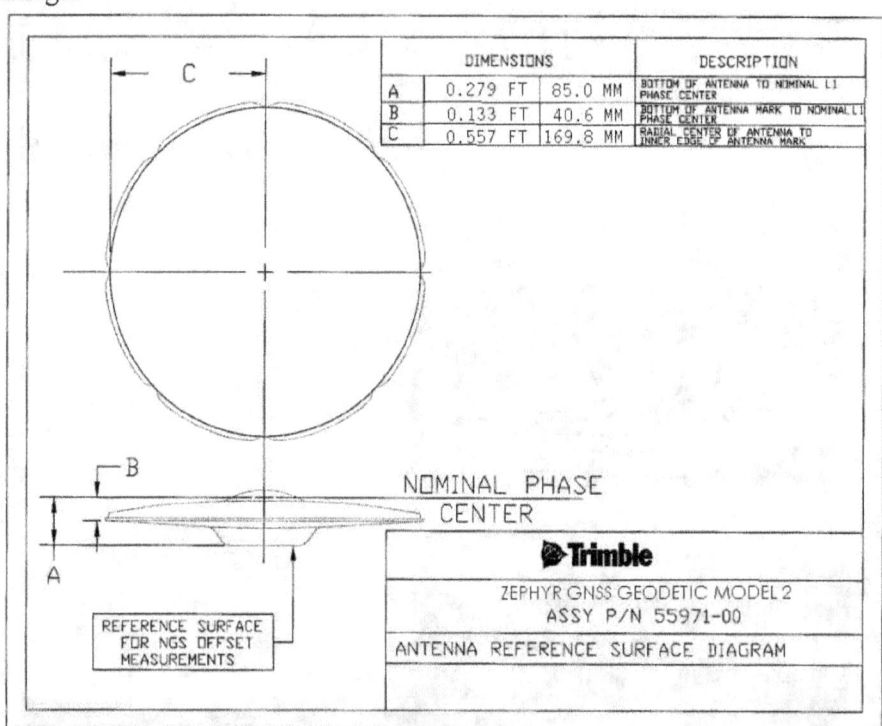

Figure 8. Dimensions of the Trimble Geodetic Model 2 antenna used at the dam crest reference station as shown on the image provided on the NGS OPS website.

These RINEX files were then submitted to National Geodetic Survey's Online Positioning User Service (OPUS http://www.ngs.noaa.gov/OPUS/) for processing to obtain the

NAD83 (CORS96 Epoch2002) NAVD88 (Geoid03) position for the benchmark on the dam crest shown in Figure 7. The antenna type and ARP height were both specified when submitting the zipped RINEX files to OPUS.

Figure 9. R. Kvitek and K. Glitz rowing from R/V MacGinitie to dam with GPS reference station for morning setup.

Figure 10. Carrying the GPS reference station to the dam crest benchmark for morning setup before start of survey.

These same reference station files and their OPUS solutions were retained for use in post-processing the vessel trajectory files used in the cleaning of the raw topographic and bathymetric survey data as described below.

3.3 Bathymetric survey

3.3.1 Acquisition

Bathymetry data was acquired with a SEA SwathPlusH 468 kHz interferometric bathymetric sidescan sonar system pole-mounted on the bow of the R/V MacGinitie (Figure 11). The SwathPlus has a 1.1° along track beam angle and achieves a 0.01 m across track resolution. SEA SwathProcessor v3 software running on a Windows XP laptop was used to record the raw sonar data.

Figure 11. R/V MacGinitie tied up to the floating dock in front of the Loch Lomond boat launch. Here the boat is configured for the bathymetric survey with the pole-mounted SwathPlus sonar head seen below the waterline at the bow. Also visible on the bow are the two Applanix POS M/V 320 GPS antennas mounted on a rigid 1.5 m long rail running fore/aft along the long axis of the boat and above the roof line to insure a clear view of the sky. Just visible at the base of the antenna mount is the edge of the yellow POS M/V IMU (inertial motion unit). The IMU, antennas and sonar head are all rigidly coupled to the same mounting plate bolted to the vessel's bow to facilitate the accurate measurement and maintenance of consistent x, y, z lever arm offsets between all sensor components of the system.

An Applanix Position and Orientation System, Marine Vessel (POS/MV 320v4) enabled for L1/L2 GPS was used to record vessel trajectory data including position and full 3D attitude (pitch, roll, yaw, and heave) data generated at 200 Hz. The POS M/V GPS antennas and IMU (inertial motion unit) were all rigidly coupled to the same mounting plate used to hold the SwathPlus sonar head pole mount insuring precise measurement and

maintenance of sensor lever arm offsets (Figure 11, 13). A Trimble DGPS Beacon receiver provided Coast Guard RTCM differential GPS corrections to the POS MV. The POS M/V system achieves an average pitch, roll and yaw accuracy of +/-0.03°. Surface-to-bottom profiles of the speed of sound through the water were collected periodically during the surveys to correct for variations in sonar beam trajectory (refraction) due to temperature and density changes throughout the water column.

Survey planning and vessel navigation were done using Coastal Oceanographics HypackMax 2008 software running on a Windows XP laptop computer with differential GPS provided via the R/V MacGinitie onboard Sinex GPS with internal DGPS beacon. Written log sheets were maintained by the crew throughout the survey recording all system configuration settings and changes, data file names, survey line start and end times, and other relevant events and comments. Log sheets were scanned and archived as digital image files. Data files were recorded to internal hard drives on the laptop computers and then transferred to external hard drives for backup and processing.

The bathymetry survey was conducted on March 28, 2009, with a patch test for system calibration conducted on March 30, 2009.

3.3.2 Post-processing

Vessel trajectory data from the files logged by the Applanix POS/MV were processed using Applanix POSPAC 5.2 software that employs a tightly coupled Inertially Aided Post Processed Kinematic (IAPPK) technique to generate a smoothed best estimate of trajectory (SBET) file at 200 Hz. The data files from the Trimble NetR5 reference station set up on the dam crest benchmark were used along with those from several other publically available continuously operating reference station (CORS) in the vicinity to create a virtual reference station (VRS) solution at the position of the survey vessel. The dam crest reference station was designated as the control station for the VRS with the NAD83 (CORS96)(epoch2002) HAE solution obtained from OPUS used as the reference station coordinates. This approach forced the VRS and subsequently generated SBET to be created in the NAD83 HAE reference frame and datum. The SBET solution includes rotational motion about all three axes as well as heave due to surface waves and water level variation over the survey period, all tied directly to the ellipsoid, virtually eliminating positional and motion-related artifacts traditionally found in multibeam data that tended to obscure fine, sub-meter geomorphic detail, particularly when data from adjacent track lines are superimposed. Applying the new IAPPK SBET approach to existing multibeam sonar data yields more co-registered data points per unit area with less noise, bringing fine features into much sharper focus than previously was possible

The raw sonar data for each survey line were then combined with the SBET and sound velocity profile data using SEA SwathProcessor software (v3.06.04.06), filtered to remove the majority of erroneous soundings, and exported in SXP format. Filters applied in SwathProcessor included those based on range, minimum and maximum depth, amplitude, angle proximity, and adaptive along-track variation. Initial processing in SwathProcessor revealed numerous issues with the existing sound velocity correction algorithms in the

software, which were reported to SEA and addressed with software updates. The version reported here is the final, updated version of the software used.

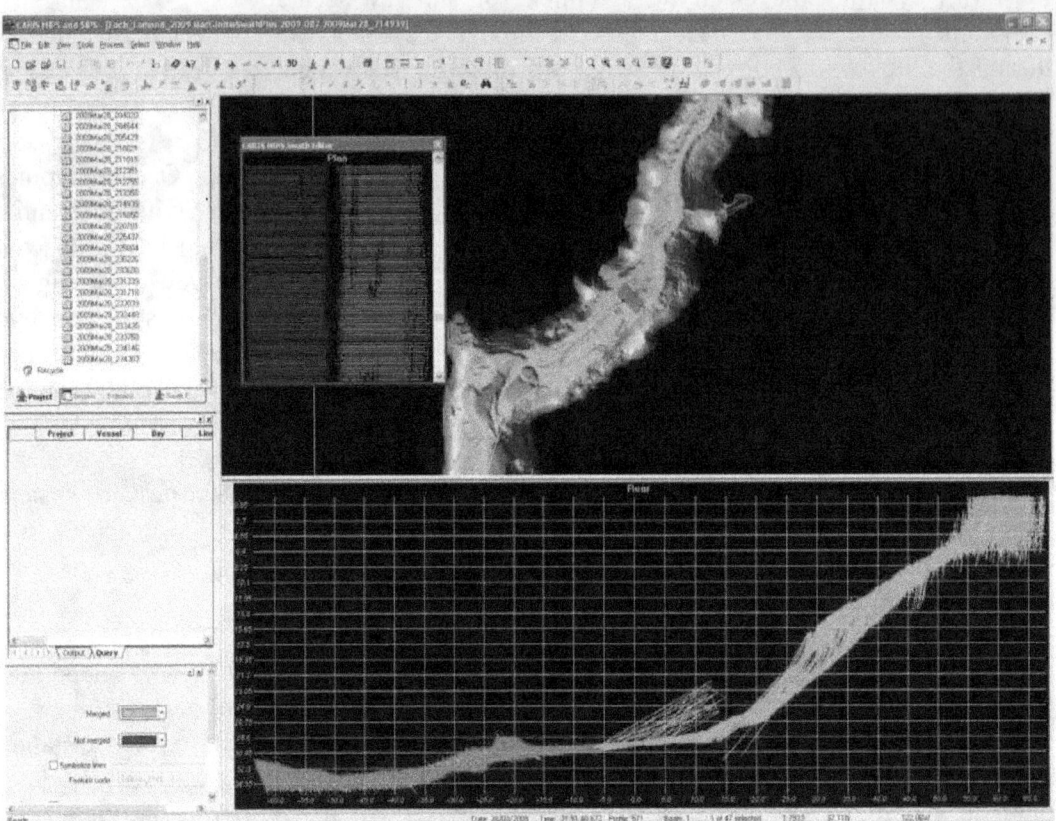

Figure 12. Bathymetric data cleaning in CARIS HIPS swath mode. Upper window shows the preliminary 0.5m CUBE surface DEM in shaded relief colored by depth. The profile window in the lower half of the screen shows the individual depth soundings from a short data segment "sliced" from the results of one survey line (yellow line in DEM). The small red/green "slice" box superimposed on the DEM shows the location and aerial extents of the profile. The Plan window is a plan view of all the data points in the red/green slice box. The profile data points are connected by lines with red = accepted points left of nadir, green = accepted points right of nadir, grey = data points rejected during cleaning process. Data processors used filters and manual techniques to identify and reject erroneous sounding values from the xyz sounding point cloud.

The SXP files were then imported and converted to HDCS format for cleaning in CARIS Hydrographic Information Processing System (HIPS) 6.1 software. Data processors using both automated filters and manual cleaning techniques examined the bathymetry point cloud looking for and eliminating erroneous data points from motion artifacts, misalignments, bottom debris, water column noise and water surface reflections. The cleaned data were then gridded at 0.5m cell size to form a CUBE surface, from which 2,815,308 xyz bathymetric point values were exported as an ascii text file as UTM zone 10 NAD83 HAE for later merging with the laser scanner topographic data. The entire cleaned bathymetry data point cloud was also exported as xyz ascii data in UTM NAD83 HAE.

3.4 Topographic survey

3.4.1 Acquisition

Topographic terrain data was acquired using a Reigl LMS-Z420i terrestrial laser scanner operated in mobile mapping mode in conjunction with the Applanix POS M/V 320 (Figure 13). The LMS-Z420i is a Class 1 laser scanner, with a maximum range of 1000 m, accuracy of 0.01m, and precision of 0.008m. The system has a vertical scan angle range of 80°, a scan rate of 8000 points/second, and an angular resolution of 0.002°.

Figure 13. Reigl LMS-Z420i terrestrial laser scanner mounted with the Applanix POS M/V IMU (yellow box) and twin GPS antennas on the R/V MacGinitie bow plate.

Equipped with the option external time sync, the scanner's internal clock is synchronized with the POS M/V GPS time using the POS PPS NMEA string and triggered via the POS generated PPS signal. The laser was controlled and its data logged using Reigl RiScanPro software running on a Windows XP laptop computer. As with the bathymetry survey, vessel trajectory data were logged using the Applanix POS M/V. The scanner recorded topographic position data points during the survey as the R/V MacGinitie was piloted around the perimeter of the reservoir shoreline, including all islands and accessible inlets. Data files were logged to the internal hard drives of the laptops and then transferred to external hard drives for back up and post-processing.

Immediately following the completion of the reservoir survey and prior to data processing, a standard patch test calibration was performed on the laser system at the CSUMB test range on March 31, 2009 to quantify any angular offsets between the IMU and laser scanner. The results from this test were then applied during the data conversion process.

3.4.2 Post-processing

Applanix PosPAC 5.2 was again used to create a SBET trajectory from the POS M/V data logged during the laser scan survey in the same manner as described in the bathymetry post-processing section above, providing a NAD83(CORS96 epoch 2002) HAE solution. Reigl POF Import was used to create a POF file for import into the RiScanPro Project. RiWorld software was then used to convert the raw scan .4dd data files, applying the patch test values for angular offsets, and combining them with the POF trajectory file to create .sdw files. RiWorld was then used to export all the uncleaned laser data as xyz topographic points in UTM zone 10 NAD83 HAE, to match coordinates of the bathymetry data.

IVS PFM Direct software was then used to create a PFM file out of the exported xyz ascii file for cleaning in Fledermaus v6.7 3D Editor. Here, the operator created a clean, bare earth data set cropped to the height of the spillway retaining wall ("y" benchmark in Figures 3 and 5) by removing all returns from vegetation, water surface debris and reflections. The remaining data points were then exported from Fledermaus as an xyz ascii file in UTM zone 10 NAD83 HAE coordinates.

3.5 *Bathy-Topo data fusion*

Before combing the topographic and bathymetric data, UltraEdit text editor was used to convert the 2.8m bathymetry soundings from negative to positive values by removing the "-" sign inserted by CARIS in the CUBE surface export. IVS PFM Direct was then used to combine the bathymetric xyz data points exported from the CARIS 0.5m CUBE surface grid with the entire cleaned, cropped bare earth topographic xyz point cloud data set exported from Fledermaus into a single PFM data file in UTM NAD83 HAE coordinates with 0.5m and 1m CUBE surfaces for further cleaning and accuracy assessment in Fledermaus 3D editor. After final cleaning, gridded xyz values for the combined data set were exported at 0.5m and 1.0m resolution from the clean bare earth PFM CUBE surfaces as UTM NAD83 HAE data points. CORPSCON 6 with Geoid03 was then used to convert these HAE points to UTM NAD83 xyz data sets as comma separated (.csv) ascii text files for three different vertical datums: NAVD88 meters, NGVD29 meters and NGVD29 feet.

3.6 *Accuracy assessment*

Accuracy assessments were conducted separately for the bathymetric and topographic survey data sets.

3.6.1 Bathymetric data accuracy assessment

The final accuracy of the cleaned and gridded bathymetric data obtained during the March 28th SwathPlus survey was assessed and verified using lead line soundings taken at precisely defined locations on March 29, 2009 (Figure 14). Lead line measurements were taken with a weighted meter tape from the top of the Applanix IMU mounted on the vessel bow (Figure 13) to the reservoir floor while the boat was tied up and held stationary at the floating dock (n = 4) and the breakwater (n = 3) outside the boat marina cove.

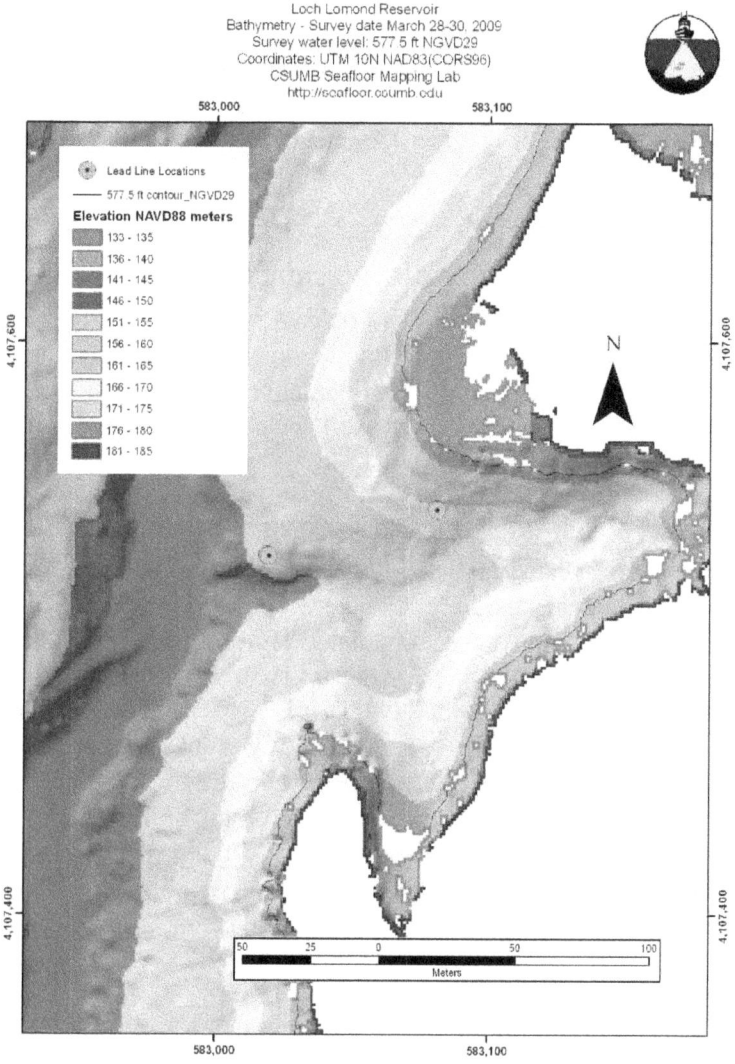

Loch Lomond Reservoir
Bathymetry - Survey date March 28-30, 2009
Survey water level: 577.5 ft NGVD29
Coordinates: UTM 10N NAD83(CORS96)
CSUMB Seafloor Mapping Lab
http://seafloor.csumb.edu

Figure 14. Locations of lead line soundings (green dots) used to assess accuracy and precision of the SwathPlus sonar bathymetry data.

Target features were marked and time-stamped in Hypack at the moment each lead line value was called out and recorded by the crew members taking the soundings, thereby recording the exact time of each sounding for later use in correlating it with the precise position and elevation of the IMU. Because the top of the IMU is the reference point for the Applanix POS M/V system, a very precise value for the IMU xyz position relative to the GPS reference station was obtained after the POS M/V data were post-processed in POSPAC in the same manner as described in the bathymetry methods section above. The UTC time of each sounding was then used to look up the xyz position of the IMU in the NAD83 HAE SBET results. The HAE value for each sounding was obtained by adding the tape measure value for the sounding to the z value of the IMU position which was located directly over the sounding location. The mean value of the sounding elevations were calculate for each of the two locations, and these were compared to: 1) the bare earth DEM grid value at the sounding location, and 2) the mean of all the accepted SwathPlus

soundings in CARIS within 0.5 m of the mean sounding location (n = 195 and 200 respectively). T-tests were then performed on the lead line versus sonar results to test for significant differences at the p = 0.05 level. The differences between the two methods and the final DEM grid values for these locations were further assessed and classified according to International Hydrographic Organization (IHO) Standards for Hydrographic Surveys Special Publication S-44:
(http://www.iho.shom.fr/publicat/free/files/S-44_5E.pdf).

3.6.2 Topographic data accuracy assessment

The accuracy of the topographic data acquired with the Reigl LMS z420i laser scanner was assessed by comparing the results to known positions of reflective targets setup over published or GPS derived vertical and/or horizontal benchmark positions and visible in the data point cloud (Figures 19 and 20) as well as objects floating at the water surface (Figure 23).

3.6.2.1 Spillway benchmark for vertical control accuracy assessment

A published benchmark located on the spillway retaining wall (Figure 15) was one of the marks employed in assessing the vertical accuracy of the topographic survey data. For this test, a 0.3m x 0.4m reflective target mounted on a leveled tripod was place above the "nail" benchmark (Figures 16-18) designated as "y" on the Bowman & Williams data sheet (Figure 15). The elevation of the "y" mark is given as 590.15 ft Santa Cruz Datum (assumed here to be NGVD29. This assumption was subsequently tested as part of the accuracy assessment.). The vertical distance between the top edge of the target and nail was measured and the target assembly left in place on the spillway while the topographic laser survey was run as described below. After the laser scanner data processing and cleaning were completed, the "y" benchmark elevation obtained from the results was compared to the published value to assess the vertical accuracy of the survey.

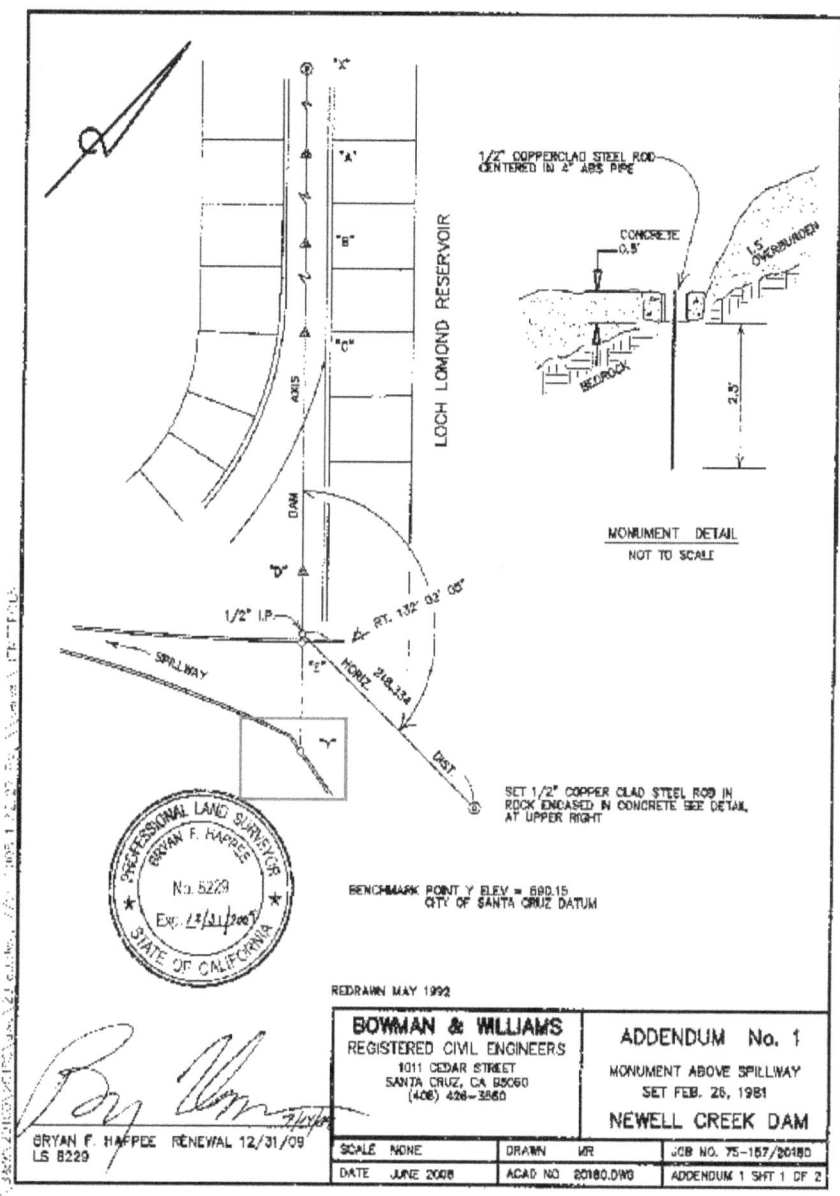

Figure 15. B&W NCD Survey data showing benchmarks along dam crest and on spillway. The spillway "y" BM nail (red box) was used to position a reflective target for calibration and validation of the laser scanner topographic survey. The elevation of the "y" mark is given as 590.15 ft Santa Cruz Datum (assumed here to be NGVD29).

Figure 16. Two views of 30x40 cm laser target set up on tripod over spillway BM nail "y" referred to in B&W data sheet shown in figure 15 above. Note the water level was at the top of the spillway.

Figure 17. Spillway BM nail "y" shown in B&W NCD Survey Data above.

Figure 18. Target set over BM "y" nail in spillway wall. Top edge of 20 x 30 cm rectangular target is 1.308m above the nail.

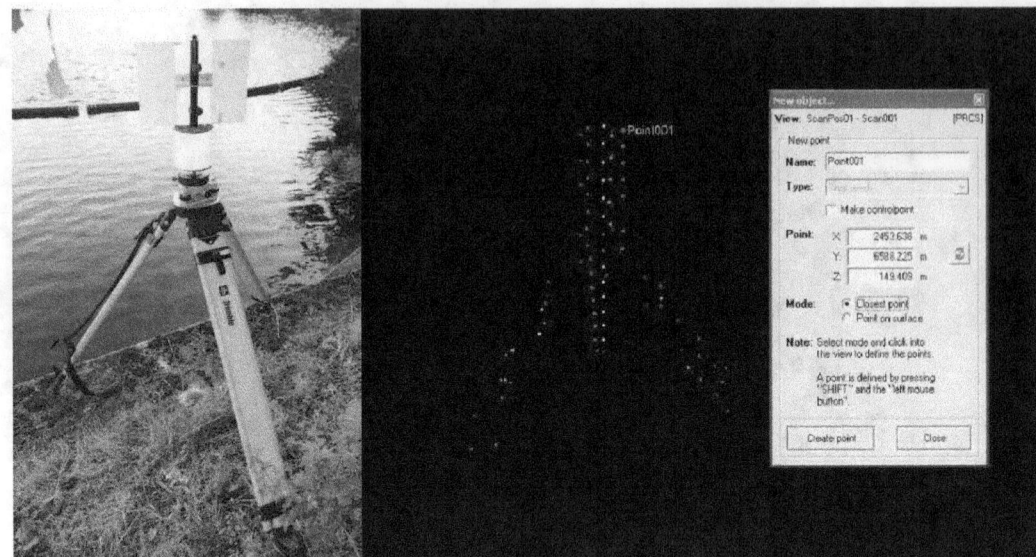

Figure 19. Laser target setup over spillway benchmark and the corresponding data points (with xyz position noted for top of reflector) in the laser data point cloud.

For the spillway benchmark comparison, the vertical value from the laser data for the top of the reflector, initially processed in NAD83 HAE, was converted to NAD83 NGVD29 using Corpscon 6 and its standard Vertcon94 data files. The measured distance of the top of the target above the spillway benchmark nail was then subtracted from the target height to obtain the benchmark height from the laser results.

3.6.2.2 Dam crest benchmark for vertical control accuracy assessment

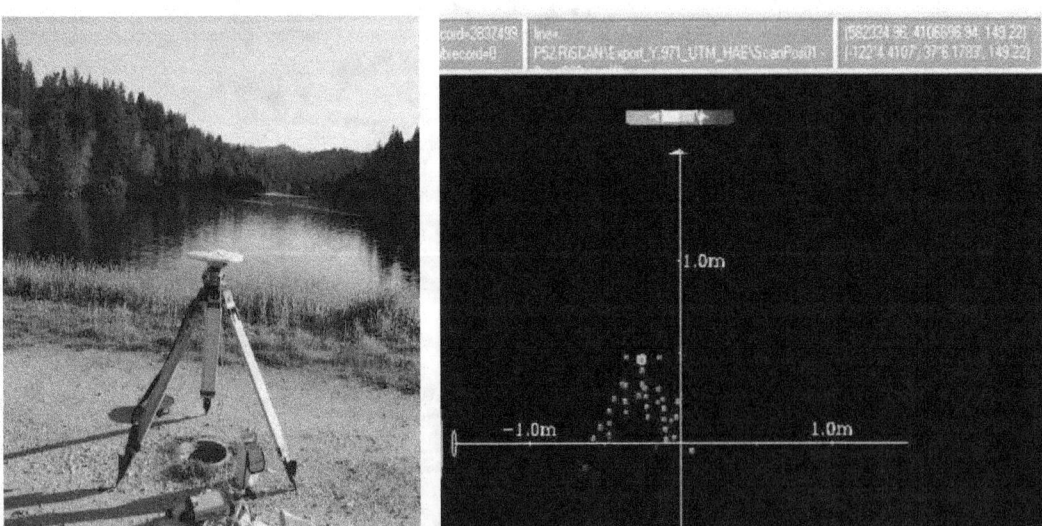

Figure 20. GPS reference station antenna set up over the dam crest benchmark and it corresponding data points as seen in the laser scanner point cloud with the xyz position of the antenna ARP (antenna reference point) selected and displayed in UTM HAE.

A similar process was used to assess the vertical and horizontal precision of the laser survey results using the dam crest benchmark position. Here, the OPUS derived antenna

reference point (ARP) solution for the GPS reference station files (described in the GPS Reference Station section above) was compared to the xyz position of the ARP identified in the laser point cloud (Figure 20).

3.6.2.3 Water level staff gauge as vertical control

There are two water level staff gauges at the reservoir, one located at the dam site just below the reference station bench mark and the other next to the boat launch (Figures 21 and 22). These staffs were used to note and record the elevation of the water level during each day's setup and breakdown of the GPS reference station, as well as opportunistically throughout the survey day. Given the correspondence between the gauge markings and the published elevation of the benchmark, it was assumed that the elevations were in NGVD29. This assumption was subsequently tested as part of the survey accuracy and precision assessment via comparison of the water level values obtained from the laser, sonar and staff gauge data.

Figure 21. Locations of reservoir water level staff gauges at the dam site (green) and boat launch (red) used for the project.

Figure 22. Reservoir water level staff gauges. Left - Staff at dam site showing water height at start of laser survey (577.3 ft. presumably NGVD29). Right – Staff at boat launch site 4 hours later.

The vertical accuracy of the topographic laser scanner survey was also assessed by comparing the water level reading on the staff gauge with the water level elevation seen in the laser data at that location within one hour of the staff reading. The laser-derived water level was calculated by taking several readings from points along the length of hose floating just at the water surface and extending out from the base of the staff gauge (Figure 23). Because the top of the hose barely broke the water surface, it provided a reliable indicator for water level elevation at the time of the laser scan.

Figure 23. Water level accuracy assessment. Left: Staff gauge used to measure water level at start of the laser survey. Note small diameter hose bending around and floating at the water surface behind the gauge. Right: The red dot is a single point selected out of many visible in the laser scanner data along the length of the floating hose. The xyz position of the selected point is shown in the inset. These laser data are from a scan run less than an hour from when the staff photo shown at left was taken. The top of the hose barely broke the water surface, providing a very accurate indicator for water level elevation in the laser data.

3.7 GIS product creation

Digital elevation models (DEM) and derivative products were built from the combined topo-bathy bare earth xyz points in ESRI ArcGIS 9.2. The .csv ascii text files were first imported into an ArcGIS project as events, and then the Spatial Analyst extension "convert features to raster" tool was used to create raster grids from the imported events. Raster grids in ArcGIS format were created in both UTM zone10 NAD83 NAVD88 and UTM zone 10 NAD83 NGVD29 coordinates to facilitate future analysis and comparisons with other data sets. Spatial analyst was also used to create contour lines and hillshades from the combined topo-bathy bare earth DEM raster grids.

4 RESULTS

4.1 Positioning control

4.1.1 GPS reference station

The post-processed solutions obtained from OPUS for the two Trimble NetR5 GPS references station files collected over the dam crest benchmark (Figure 7) were highly consistent. Differences between the UTM coordinate solutions for the two days were x = 0.008m, y = 0.005 m and z = 0.011 m (Figures 24 and 25).

Figure 24. NGS OPUS solution for Static GPS data collect at dam BM March 28, 2009.

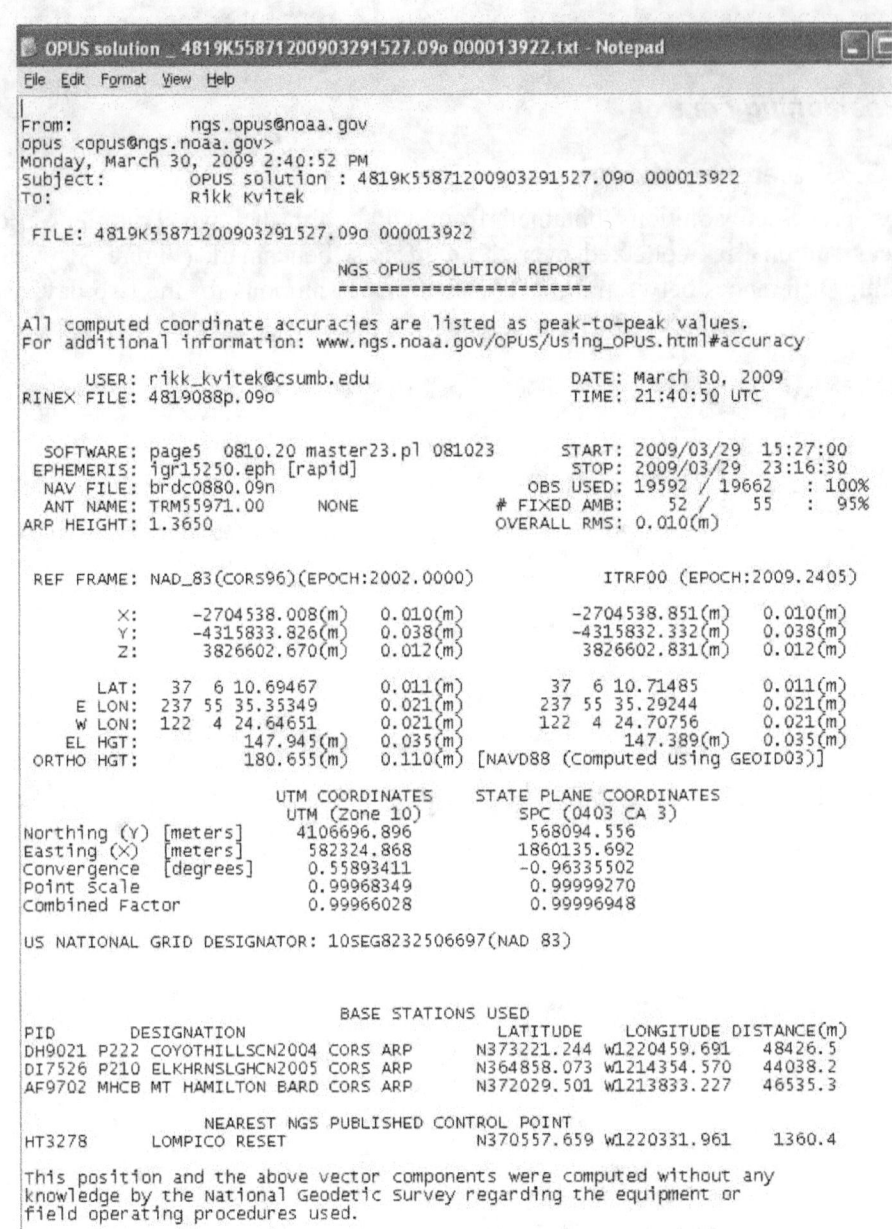

Figure 25. NGS OPUS solution for Static GPS data collect at dam BM March 29, 2009.

The following NAD83(CORS96)(EPOCH:2002.0000) reference frame position values obtained from OPUS for the March 29, 2009 data file were arbitrarily selected as the ones used for the dam crest benchmark in all subsequent processing and analyses including use as the control station coordinates for processing all vessel trajectory files in POSPAC.

Latitude	37 6 10.69467
West Longitude	122 4 24.64651
Elevation (HAE)	147.945(m)
Elevation (NAVD88)	180.655(m) (Computed using GEOID03)

4.2 Bathymetric survey

A total of 145,692,732 depth soundings were exported as SXP files from SwathProc after initial filtering (Figure 26). Following import of the SXP files into CARIS, 19,054,590 soundings were rejected by the data processors, leaving a total of 126,638,142 cleaned soundings in the xyz bathymetry point cloud. These remaining points were used to create the final 0.5m CARIS CUBE surface consisting of 2,815,308 cleaned bathymetric values at a grid spacing of 0.5m covering 173.9 acres of the wetted reservoir basin up to the 577.5 ft waterline (Figure 27). This planar extent closely matches the 175 acre reservoir size published by the Santa Cruz City Water Department on their website: http://www.ci.santa-cruz.ca.us/wt/recreation/index.html.

Figure 26. The 145,692,732 SwathPlus sonar bathymetry depth soundings after preliminary filtering in SwathProc displayed here in shaded relief and colored by depth in a GoogleEarth screen capture. A total of 126,638,142 accepted soundings remained in the xyz point cloud after final cleaning in CARIS.

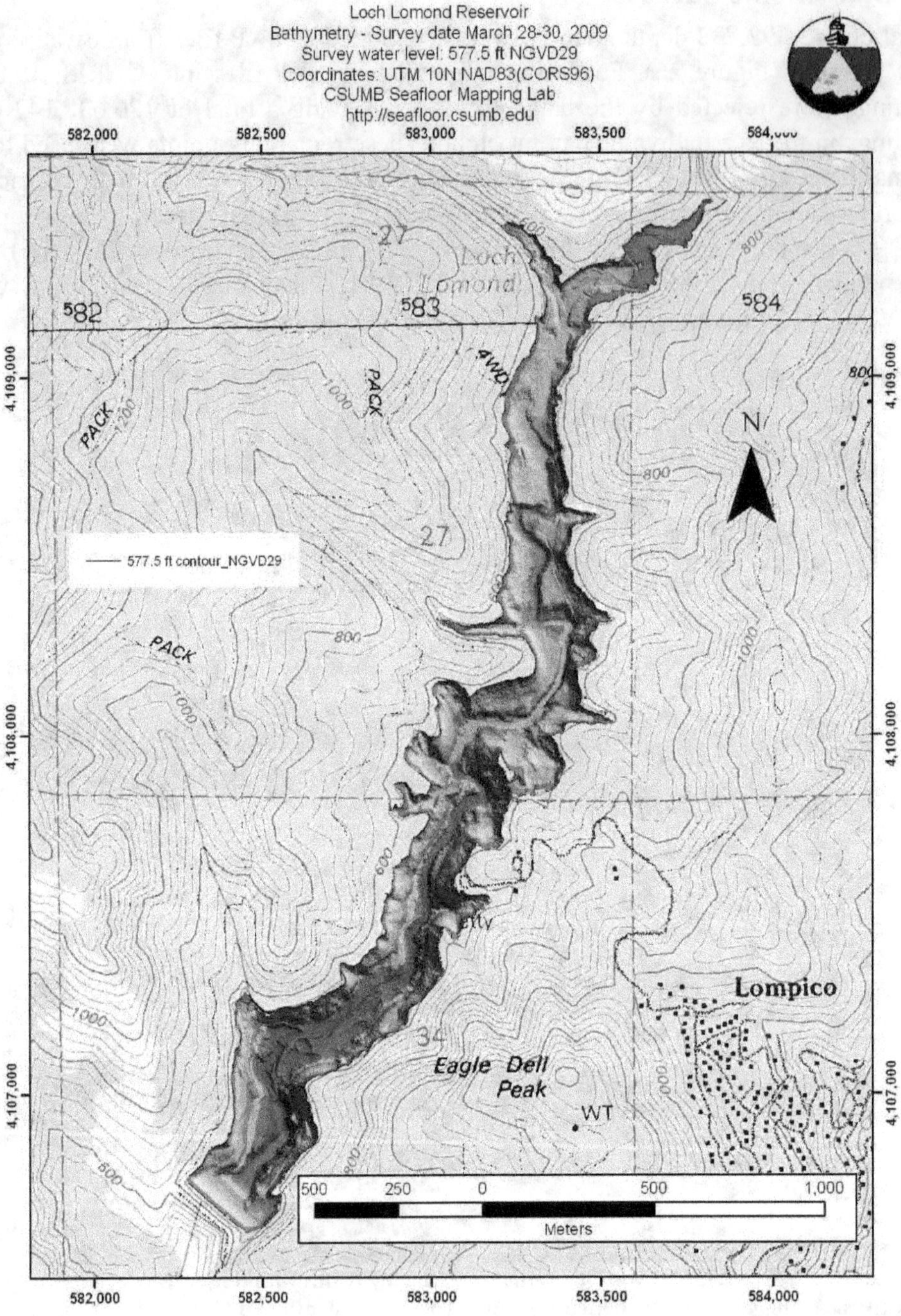

Loch Lomond Reservoir
Bathymetry - Survey date March 28-30, 2009
Survey water level: 577.5 ft NGVD29
Coordinates: UTM 10N NAD83(CORS96)
CSUMB Seafloor Mapping Lab
http://seafloor.csumb.edu

Figure 27. Coverage of bathymetry data colored by depth and shown in shaded relief. The 0.5m bathymetry DEM contains over 2.8 million cell values.

4.3 Topographic survey

A total of 3,215,710 clean bare earth topographic points were created from the laser scanner data covering the exposed basin terrain from the 577.5 ft NGVD29 water level to the 590.49 ft NGVD29 top of the spillway retaining wall (Figure 28).

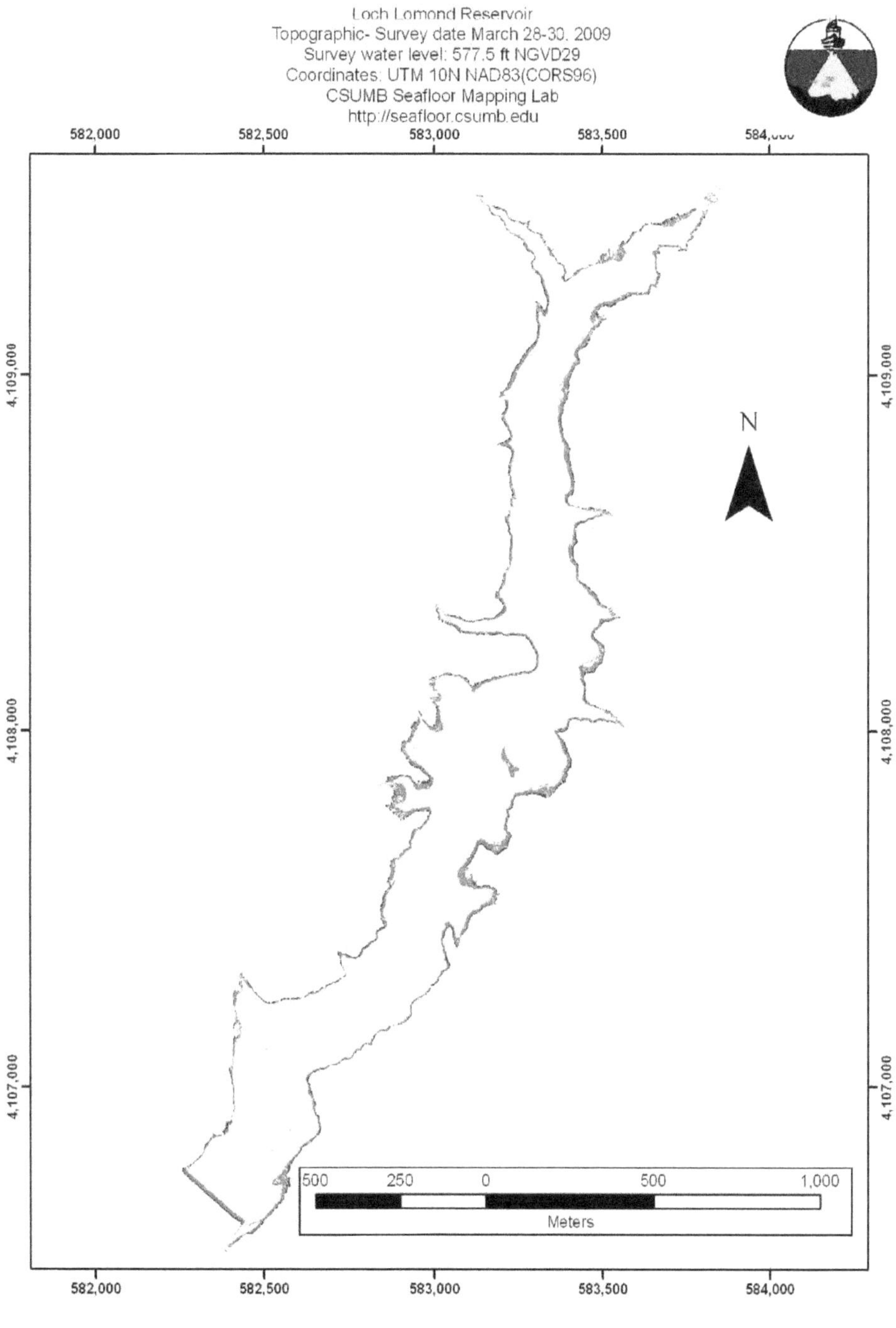

Figure 28. Aerial extent of the topographic laser survey bare earth results above the 577.5 ft 590.49 ft NGVD29 water surface and cropped to the top of the spillway retaining wall 590.49 ft NGVD29 elevation.

4.4 Bathy-Topo data fusion

The cleaned bare earth bathymetric and topographic point cloud data sets were combined to create gridded xyz data sets a 0.5 and 1.0 meter resolution for the entire basin containing 3,0634,431 and 794,759 values respectively. The 1.0m gridded data were then used to create a 1m DEM containing a total of 794,759 cell values covering the reservoir basin from the floor up to the 590.49 ft NGVD29 top of the spillway retaining wall (Figure 29).

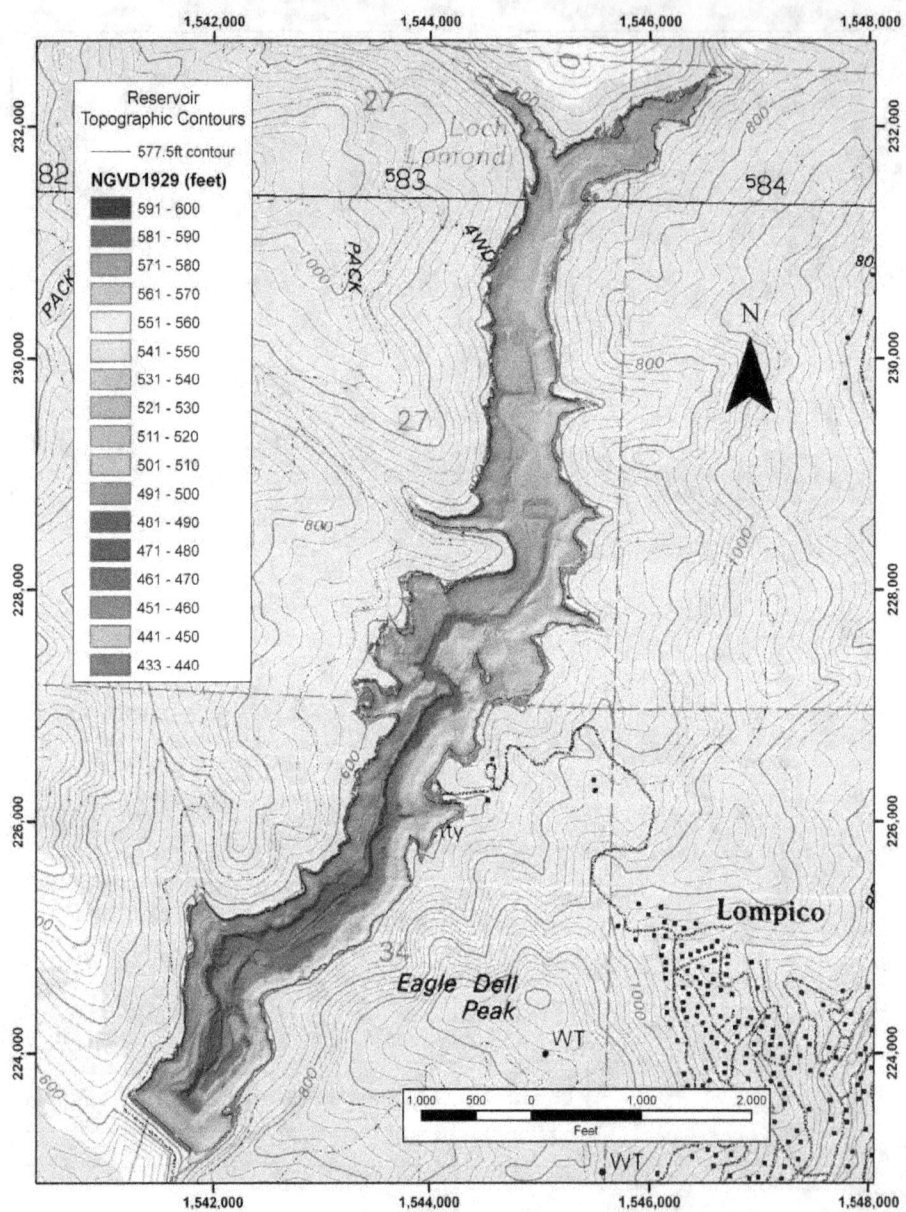

Figure 29. Combined bathymetric/topographic 1m bare earth DEM of Loch Lomond Reservoir from the basin floor to the 590.49 ft NGVD29 top of the spillway retaining wall shown in shaded relief and colored by depth.

4.5 Accuracy assessment

4.5.1 Bathymetric data accuracy assessment

The accuracy assessment indicated that the survey results met the tolerances for IHO Special Order (Table 1), the highest accuracy class specified by the International Hydrographic Organization (IHO) Standards for Hydrographic Surveys Special Publication S-44: (http://www.iho.shom.fr/publicat/free/files/S-44_5E.pdf). There was no significant difference found between the bathymetric depths obtained from the lead line test soundings and sonar results at each test site (two sample t-test @ $p = 0.05$).

Table 1. Depths in height above ellipsoid (HAE meters) obtained at the two lead line sampling sites (Figure 19) from the lead line, cleaned sonar xyz soundings and 1m final DEM product. Two-sample t-tests of the data found no significant difference between the lead line and sonar derived means at the $p = 0.05$ level. Both the lead line versus sonar and lead line versus DEM differences fall with the specification tolerances for IHO Special Order calculated for the depth at each test site.

Test site	Lead line depth(HAE) mean ± SE, n	Sonar depth (HAE) mean ± SE, n	DEM depth (m)	Lead line vs. Sonar (m)	Lead line vs. DEM (m)	IHO SO (m)
1	136.33 ± 0.016, 4	136.50 ± 0.016, 200	136.58	-0.17	-0.25	± 0.26
2	119.71 ± 0.071, 3	120.02 ± 0.014, 195	119.84	-0.32	-0.13	± 0.32

4.5.2 Topographic data accuracy assessment

4.5.2.1 Survey vertical control accuracy assessment

Verification and quantification of the GPS Vertical Control at the benchmark on the Loch Lomond Dam Crest (Figure 7) showed a high degree of agreement between measurement methods (Table 2). The benchmark elevation derived from the laser scanner topographic survey results were within 0.09 m of the static GPS solution obtained from OPUS for the March 29, 2009 7.75 hour reference station data file.

Table 2. Results for dam crest benchmark elevation and topographic survey vertical accuracy assessment (Figures 7 and 20). Shown are the NAD83 (CORS96)(epoch:2002.0000) dam crest benchmark elevation values obtained from the NGS OPUS static GPS solution for the March 29, 2009 reference station file versus the value obtained from the laser scanner point cloud (Figure 20). The values given are for the GPS Antenna Reference Point (ARP) minus the antenna height above the benchmark as measured in the field (1.365 m). Elevations are presented in height above the ellipsoid (HAE), NAVD88 (Geoid03) and NGVD29 (Vertcon94). Corpscon 6 was used for datum conversion.

Source	HAE (m)	NAVD88 (m)	NGVD29 (ft)
NGS OPUS static GPS Solution	147.945	180.655	589.97
Laser scanner	147.855	180.565	589.68
Difference	0.090	0.090	0.29

4.5.2.2 Spillway benchmark for vertical control accuracy assessment

The small difference (0.10 m) between the elevation values obtained for the Loch Lomond Spillway benchmark "Y" with the laser scanner (Figure 19) versus those published in the 1981 Bowman and Williams data sheet (Figure 15) suggests that the City of Santa Cruz Datum listed on the data sheet is indeed NGVD29 (Table 3).

Table 3. Results from spillway benchmark topographic vertical accuracy assessment (Figures 15-19). Elevations are for the "Y" benchmark shown on the Bowman and Williams 1981 data sheet (Figure 15) obtained from the data sheet versus those derived from the topographic laser survey results (Figure 19). The laser results are adjusted for the 1.308 m target elevation above the benchmark.

Source	Elevation (ft)	Datum
B&W 1981 data sheet	590.15	City of Santa Cruz Datum
Laser scanner	590.49	NGVD29
Difference	0.34	

4.5.2.3 Water level staff gauge as vertical control

The water level determined from the laser scanner data after conversion to NGVD29 was also in close agreement with the staff gauge value read at the same time (Figure 23). This small difference (0.2ft) suggests that the Santa Cruz Datum is indeed NGVD29 (Table 4).

Table 4. Water level staff gauge accuracy assessment of topographic laser survey results. The staff gauge reading was taken less than one hour before the laser points were collected on March 29, 2009.

Source	Elevation (ft)	Datum
Staff gauge	577.3	City of Santa Cruz Datum
Laser scanner	577.5	NGVD29
Difference	0.2	

4.6 Data products

A variety of point, vector and raster GIS data products including FGDC metadata were created from the survey results over and above those required in the statement of work. These files, created in UTM and State Plane coordinates and NAVD88, HAE and NGVD29 datums, are listed in Table 5 and have been provided to the USGS.

Table 5. Survey Data Products. Listing of all major data product files created from the survey results including data format, type, file name, resolution, coordinates and datums, and a description of the source and process where applicable.

Data format	Data type	Data file	Resolution	Coordinates and Datum	Description
xyz ASCII	Bathy	BathyPointCloud_xy-z_UTM_HAE.csv	Full	UTM NAD83 HAE	Caris export of cleaned Swathplus data from 0.5m cube surface. Soundings as negative. Processor: Kate Thomas
xyz ASCII	Bathy	BathyPointCloud_xy+z_UTM_HAE.csv	Full	UTM NAD83 HAE	Caris export of cleaned Swathplus data from 0.5m cube surface. Soundings as positive.

xyz ASCII	Bathy	LochLomond_50cm_xy-z.txt	0.5 m	UTM NAD83 HAE	Processor: Kate Thomas
xyz ASCII	Bathy	LochLomond_50cm_xy-z.txt	0.5 m	UTM NAD83 HAE	Caris export of cleaned Swathplus data from 0.5m cube surface. Soundings as negative. Processor: Kate Thomas
xyz ASCII	Bathy	LochLomond_50cm_xy+z.txt	0.5 m	UTM NAD83 HAE	Caris export of cleaned Swathplus data from 0.5m cube surface. Soundings as positive. Processor: Kate Thomas
xyz ASCII	Topo	LL_laser_UTM_HAE_BE.xyz	Full	UTM NAD83 HAE	Reigl export of full resolution bare earth between waterline and level of spillway retaining wall. Cleaning: All vegetation removed. All points below waterline and above top of spillway retaining wall removed. Processor: Steven Quan
xyz ASCII	Topo/bathy	BE_TopoBathy1m_NAD83_UTM_NAVD88.csv	1.0 m	UTM NAD83 NAVD88 (Geoid03)	PFM/Fledermaus export of combined 0.5 m Bathy and full resolution bare earth Topo files from 1m CUBE at IHO Special Order. Corpscon: xyz exports converted to NAVD88 Processor: Rikk Kvitek
xyz ASCII	Topo/bathy	BE_TopoBathy1m_NAD83_UTM_NGVD29m.csv	1.0 m	UTM NAD83 NGVD29 meters (Vertcon94)	PFM/Fledermaus export of combined 0.5 m Bathy and full resolution bare earth Topo files from 1m CUBE at IHO Special Order. Corpscon: xyz exports converted to NGVD29 Processor: Rikk Kvitek
xyz ASCII	Topo/bathy	BE_TopoBathy1m_NAD83_UTM_NGVD29ft.csv	1.0 m	UTM NAD83 NGVD29 feet (Vertcon94)	PFM/Fledermaus export of combined 0.5 m Bathy and full resolution bare earth Topo files from 1m CUBE at IHO Special Order. Corpscon: xyz exports converted to NGVD29 Processor: Rikk Kvitek
xyz ASCII	Topo/bathy	BE_TopoBathy50cm_NAD83UTMNAVD88.csv	0.5 m	UTM NAD83 NAVD88 (Geoid03)	PFM/Fledermaus export of combined 0.5 m Bathy and full resolution bare earth Topo files from 0.5m CUBE at IHO Special Order. Corpscon: xyz exports converted to NAVD88 Processor: Rikk Kvitek
xyz ASCII	Topo/bathy	BE_TopoBathy50cm_NAD83UTMNGVD29m.csv	0.5 m	UTM NAD83 NGVD29 meters	PFM/Fledermaus export of combined 0.5 m Bathy and full

				(Vertcon94)	resolution bare earth Topo files from 0.5m CUBE at IHO Special Order. Corpscon: xyz exports converted to NGVD29 Processor: Rikk Kvitek
xyz ASCII	Topo/bathy	BE_TopoBathy50cm_NAD83UTMNGVD29ft.csv	0.5 m	UTM NAD83 NGVD29 feet (Vertcon94)	PFM/Fledermaus export of combined 0.5 m Bathy and full resolution bare earth Topo files from 0.5m CUBE at IHO Special Order. Corpscon: xyz exports converted to NGVD29 Processor: Rikk Kvitek
DEM	Topo/bathy	be_utm_navd88	1 m	UTM NAD83 NAVD88 (Geoid03)	Bare earth Digital Elevation Models (DEM) in ArcGIS raster grid format
DEM	Topo/bathy	be_utm_ngvd29	1 m	UTM NAD83 NGVD29 (Vertcon94)	Bare earth Digital Elevation Models (DEM) in ArcGIS raster grid format
DEM	Topo/bathy	be_sp_ngvd29	1 m	State Plane NAD27 NGVD29	Bare earth Digital Elevation Models (DEM) in ArcGIS raster grid format
Hillshade	Topo/bathy	hs_be_1x	1m	UTM NAD83 NAVD88 (Geoid03)	Hillshade of bare earth merged data at 1x exageration
Geotiff	Bathy	LochLomond_1m_GS.tif	1m	UTM NAD83	Hillshade of bathymetry only in gray scale
Geotiff	Bathy	LochLomond_1m_10clr.tif	1m	UTM NAD83	Hillshade of bathymetry only colored by depth
Geotiff	Topo/Bathy	BE_LochLomondSP_HS_gs1m.tif	1m	State Plane	Hillshade of bare earth topo/bathy DEM in gray scale
Geotiff	Topo/Bathy	BE_LochLomondSP10ftcntr.tif	1m	State Plane	Hillshade of bare earth topo/bathy DEM colored by depth
Shapefile	Topo/bathy contour lines	Contours_2m_NAD83_NAVD88.shp		UTM NAD83	DEM derived contours at 2 meter intervals
Shapefile	Topo/bathy contour lines	577-5ft_contour_NGVD29.shp		UTM NGVD29	577.5 ft contour (water level at time of survey)
Shapefile	Topo/bathy contour lines	Contours_UTM_NGVD29_10ft.shp		UTM NGVD29	DEM derived contours at 10 ft intervals
Shapefile	Topo/bathy contour lines	Contours_UTM_NGVD29_40ft.shp		UTM NGVD29	DEM derived contours at 40 ft intervals
Shapefile	Topo/bathy contour lines	577-5ft_contour.shp		State Plane NGVD29	DEM derived contours at 577.5ft level (water level at time of survey)
Shapefile	Topo/bathy contour lines	590ft_contour.shp		State Plane NGVD29	DEM derived contours at 590 ft level (top of dam spillway wall and benchmark)
Shapefile	Topo/bathy contour lines	Contours_SP_NGVD29_10ft.shp		State Plane NGVD29	DEM derived contours at 10 ft intervals
Shapefile	Topo/bathy contour lines	Contours_SP_NGVD29_40ft.shp		State Plane NGVD29	DEM derived contours at 40 ft intervals

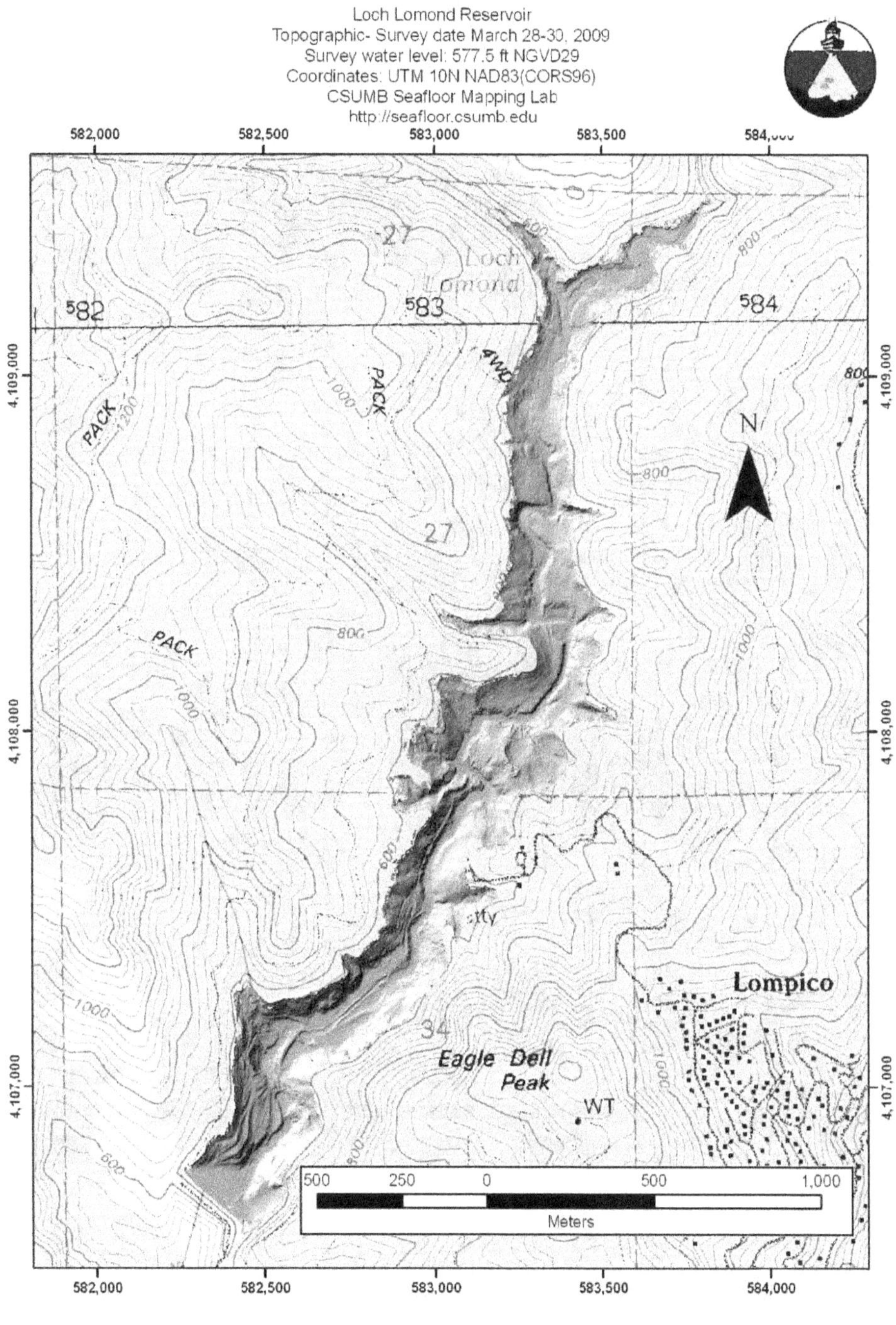

Loch Lomond Reservoir
Topographic- Survey date March 28-30, 2009
Survey water level: 577.5 ft NGVD29
Coordinates: UTM 10N NAD83(CORS96)
CSUMB Seafloor Mapping Lab
http://seafloor.csumb.edu

Figure 30. Bare earth DEM of the Loch Lomond reservoir basin shown in gray scale shaded relief.

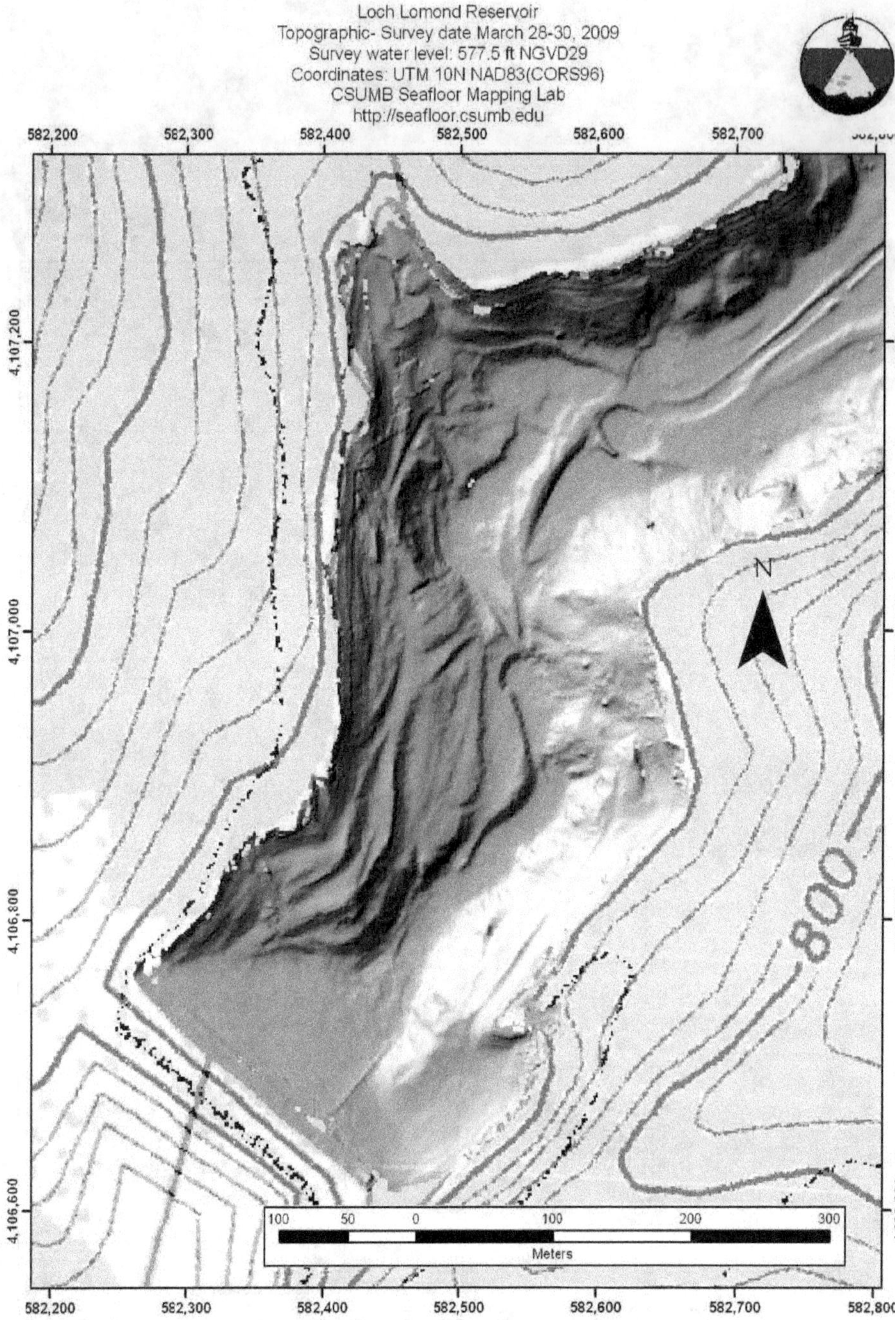

Loch Lomond Reservoir
Topographic- Survey date March 28-30, 2009
Survey water level: 577.5 ft NGVD29
Coordinates: UTM 10N NAD83(CORS96)
CSUMB Seafloor Mapping Lab
http://seafloor.csumb.edu

Figure 31. Close-up view of 1m DEM in shaded relief at the dam site showing relict features in fine detail visible on the reservoir floor.

5 CONCLUSION

The novel approach employed by the CSUMB Seafloor Mapping Lab using mobile vessel-mounted sonar and LiDAR for full-basin reservoir mapping proved highly successful. The interferometric sidescan sonar system was able to map bathymetry from the reservoir floor up to the water surface, and the mobile laser scanner covered the exposed basin topography from the water surface up to the top of the spillway retaining wall. Manual and automated cleaning and filtering were able to remove all water column debris and terrestrial vegetation yielding a cleaned, comprehensive, high-density merged bathy/topo xyz point cloud containing > 130 million individual data points. These points were used to create final cleaned bare earth gridded xyz data sets of the entire 175 acre basin at 1.0 and 0.5 m resolution that met or exceeded IHO Special Order standards for hydrographic surveys.

Appendix B.

Particle-size data for bed sediment samples collected at Loch Lomond Reservoir, Santa Cruz County, California, May 19, 2009.

Appendix B.

Particle-size data for bed sediment samples collected at Loch Lomond Reservoir, Santa Cruz County, California, May 19, 2009.

SL DS Se iment La orator   n ironmental Data S stem **Abbreviations:** mm millimeters   ercent   less t an

SLEDS sample identifier	Sample location (see fig. 9)		Time	% <8.0 mm	% <4.0 mm	% <2.0 mm	% <1.0 mm	% <0.50 mm	% <0.25 mm	% <0.125 mm
A	ange									
A	ange									
A	ange									
A	ange									
A	ange									
A	ange									
A	ange									
A	ange									
A	ange									
A	ange									
A	ange									
A	ange									
A	ange									
A	ange									
A	ange									
A	ange									
A	ange									
A	ange	L								
A	ange	L								
A	ange									
A	ange									
A	ange									
A	ange	L								
A	ange	L								
A	ange									
A	ange									
A	ange									
A	ange	L								
A	ange	L								
A	ange									
A	ange									
A	ange	L								
A	ange									
A	ange									
A	ange									
A	ange A									

Appendix B.

Particle-size data for bed sediment samples collected at Loch Lomond Reservoir, Santa Cruz County, California, May 19, 2009.—Continued

SL DS Se iment La orator n ironmental Data S stem **Abbreviations:** mm millimeters   ercent   less t an

SLEDS sample identifier	Sample location (see fig. 9)		Time	% <0.063 mm	% <0.031 mm	% <0.016 mm	% <0.008 mm	% <0.004 mm	% <2 microns
A	ange								
A	ange								
A	ange								
A	ange								
A	ange								
A	ange								
A	ange								
A	ange								
A	ange								
A	ange								
A	ange								
A	ange								
A	ange								
A	ange								
A	ange								
A	ange								
A	ange	L							
A	ange	L							
A	ange								
A	ange								
A	ange								
A	ange	L							
A	ange	L							
A	ange								
A	ange								
A	ange								
A	ange	L							
A	ange	L							
A	ange								
A	ange								
A	ange	L							
A	ange								
A	ange								
A	ange								
A	ange A								

Appendix C.

Transect profiles showing changes in reservoir bed altitude between investigations of Loch Lomond Reservoir, Santa Cruz County, California.

Appendix C.

Transect profiles showing changes in reservoir bed altitude between investigations of Loch Lomond Reservoir, Santa Cruz County, California.

egra s elo so t e altit eso t ereser oir e s r ace or t e ario ss r e so Loc Lomon eser oir   an   ra s labeled "1982" show profiles designated with the 1982 transect numbers and measured at the 1982 transect locations (shown in figure 5). Data are available for these transects from the 1960, 1982, 1988, and 2009 investigations. Graphs labeled "1971" show profiles designated with the 1971 transect numbers and mea sured at the 1971 transect locations (also shown in figure 5). Data are available for these transects from the 1960, 1971, 1998, and 2009 investigations. Profiles not incl e ere etermine to e incorrectl locate

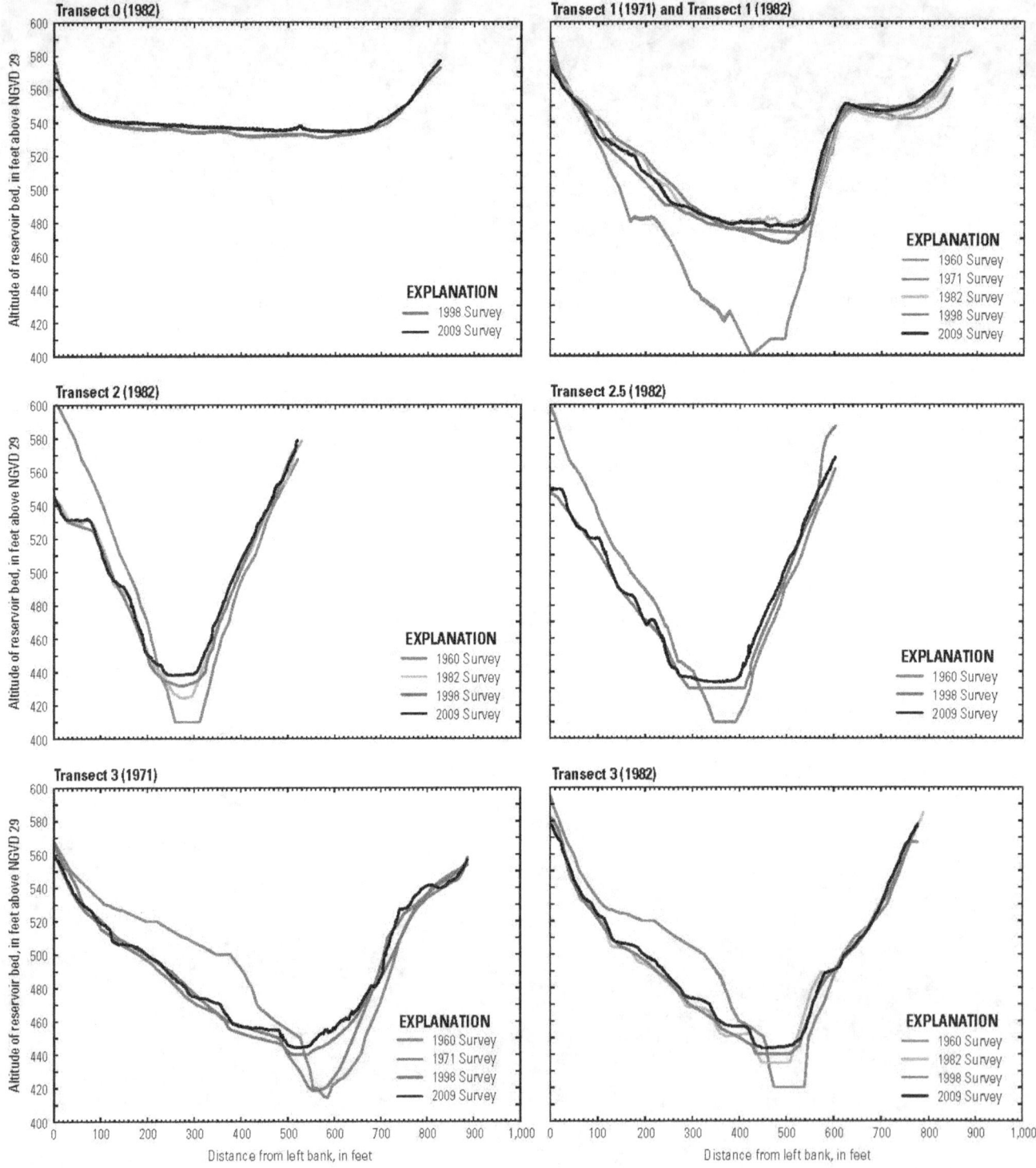

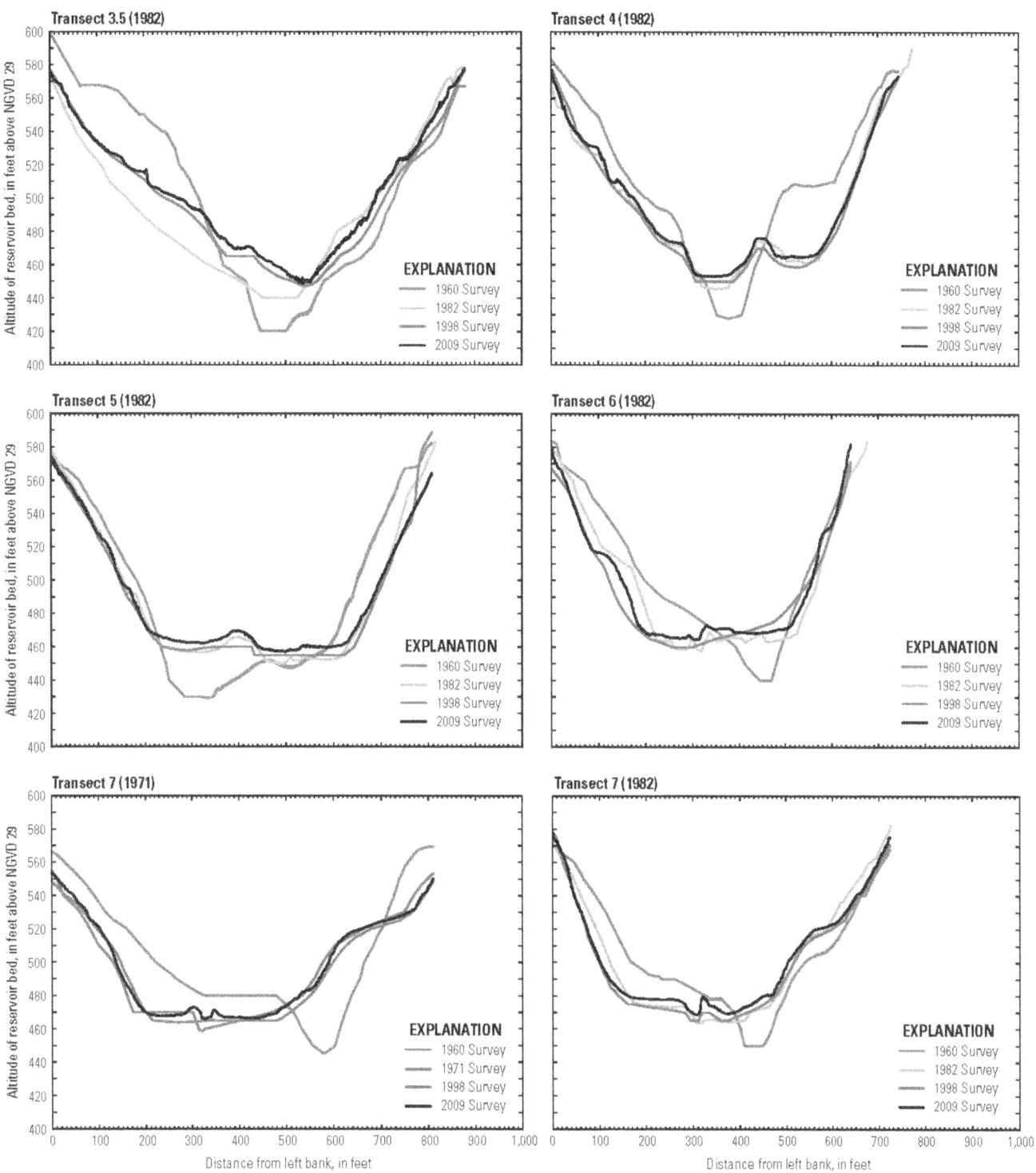

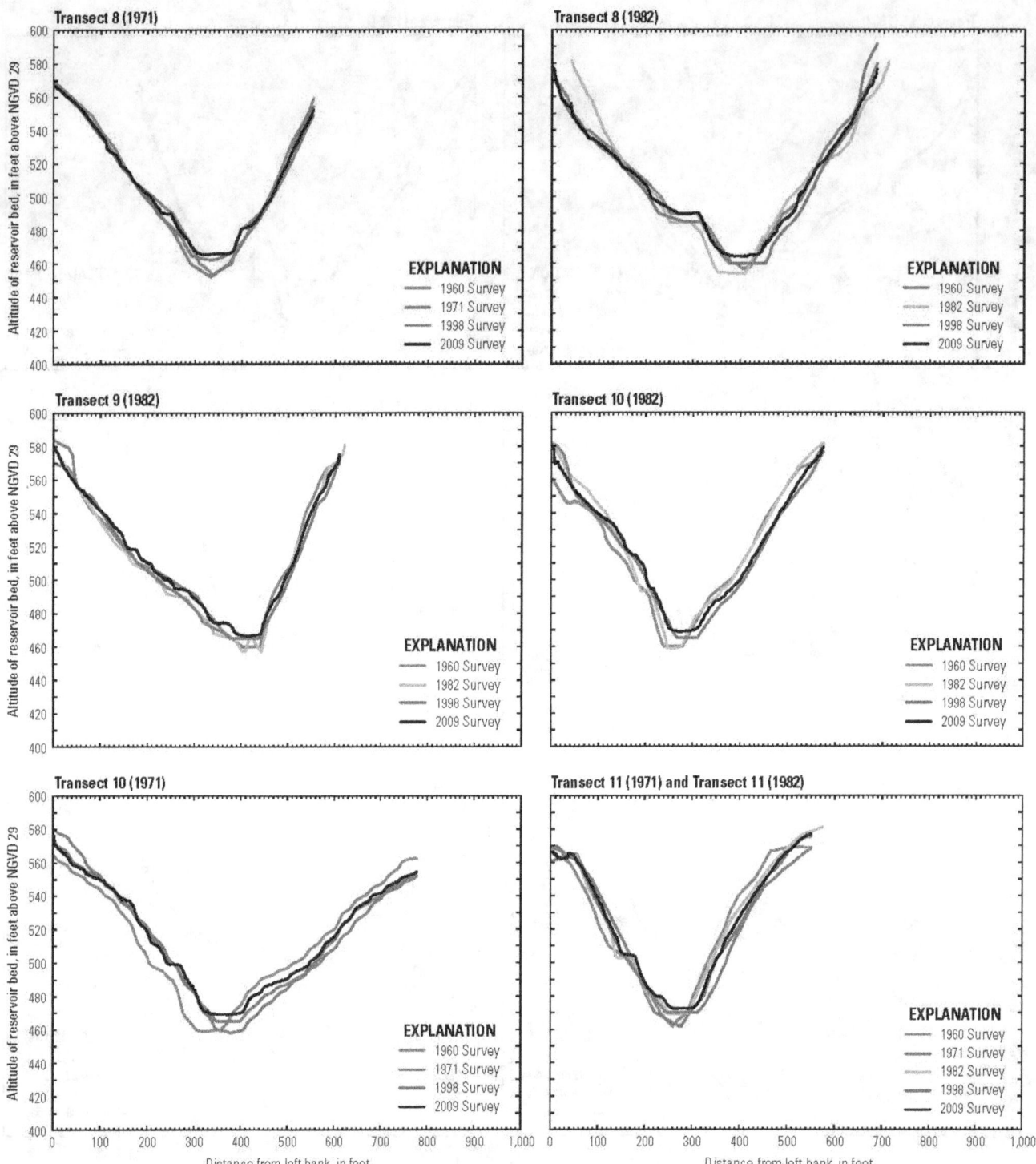

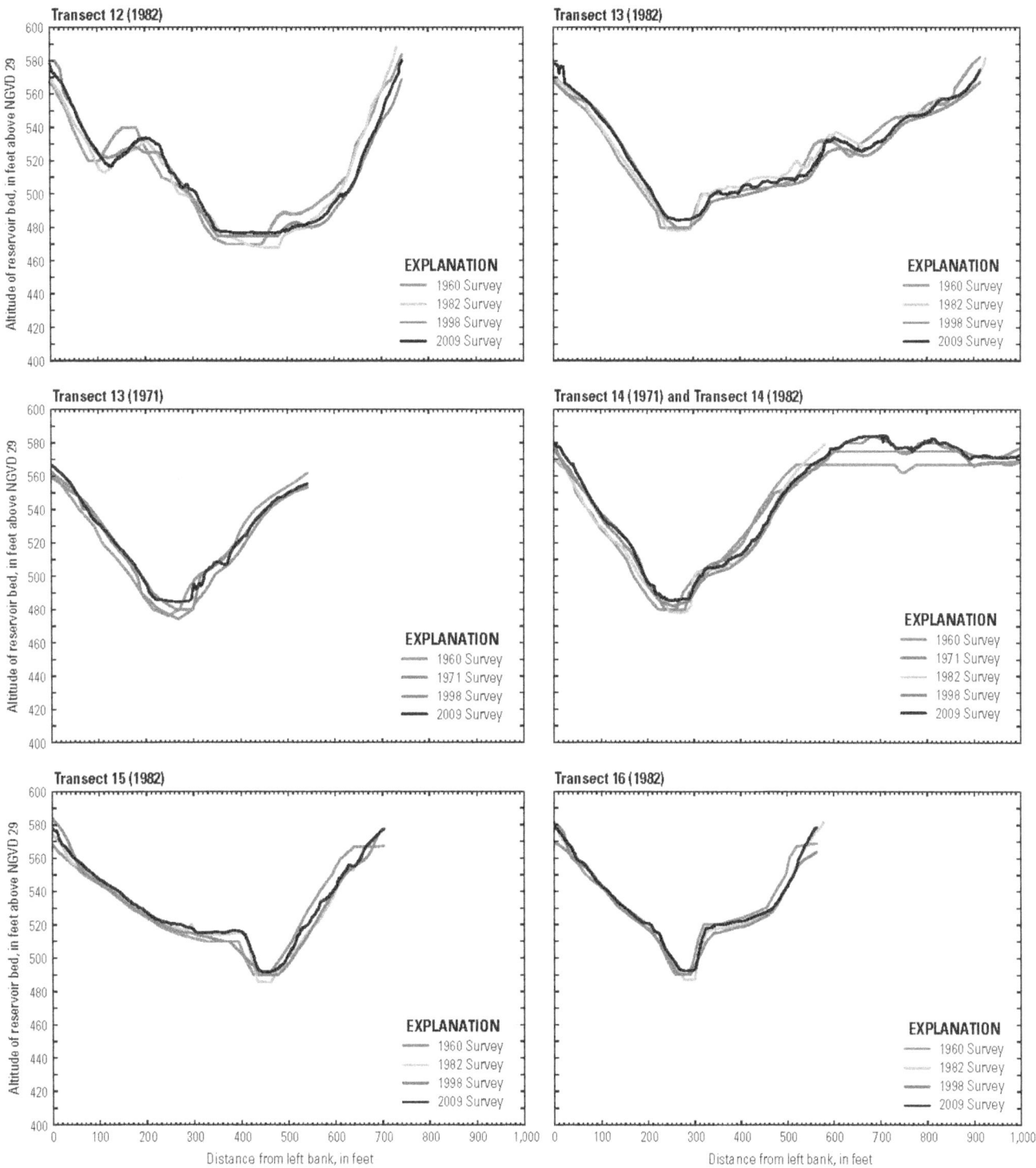

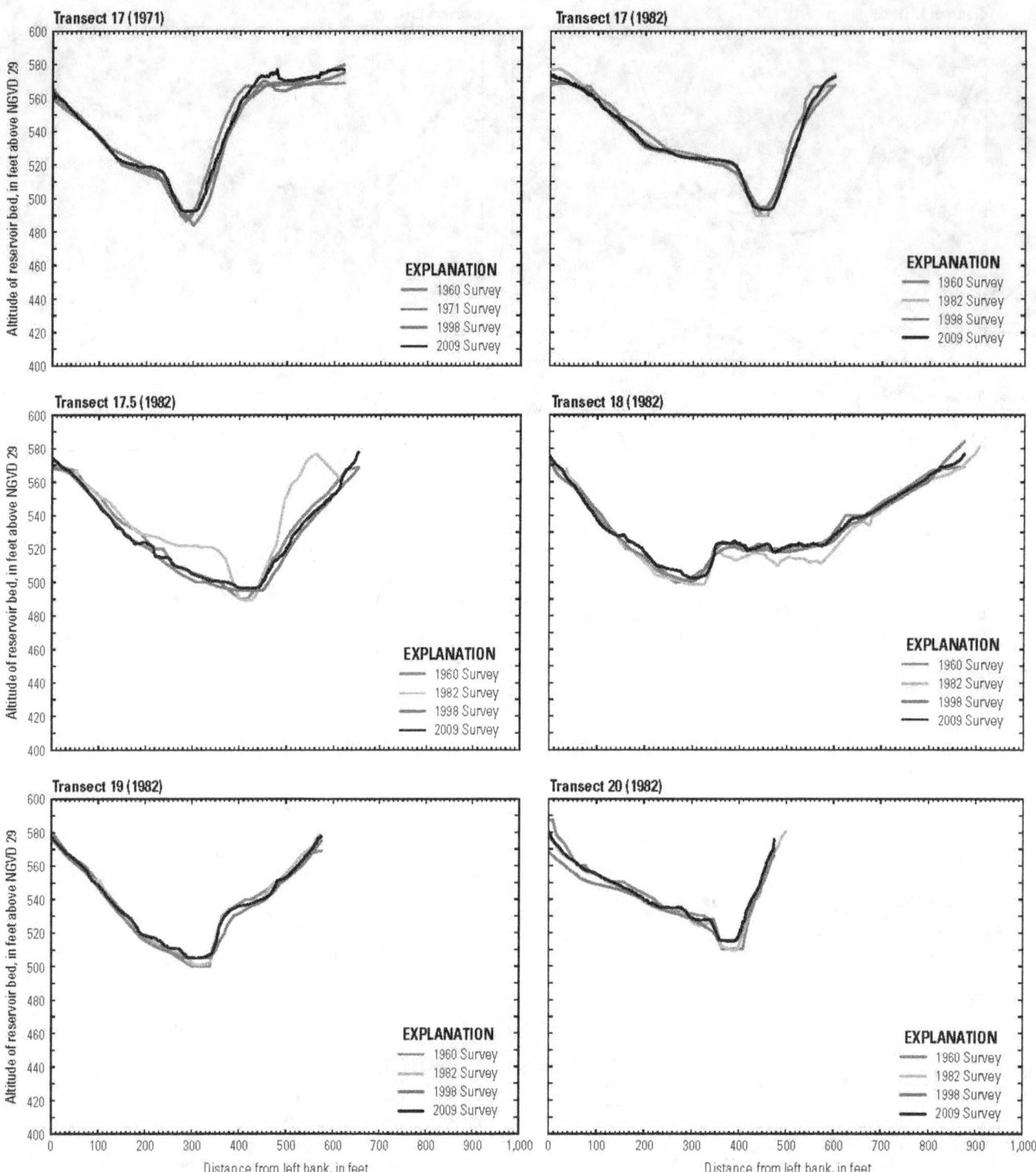

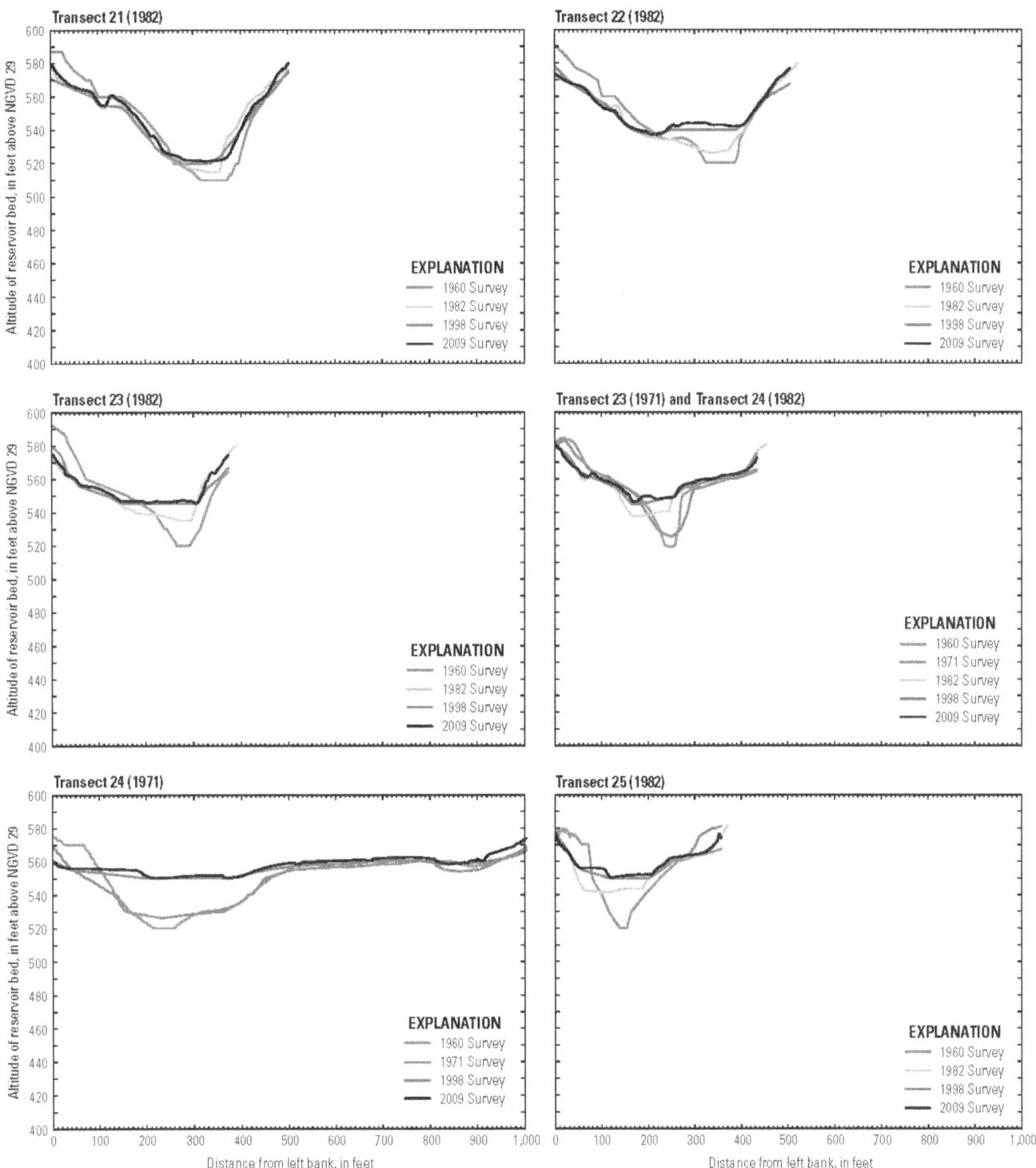

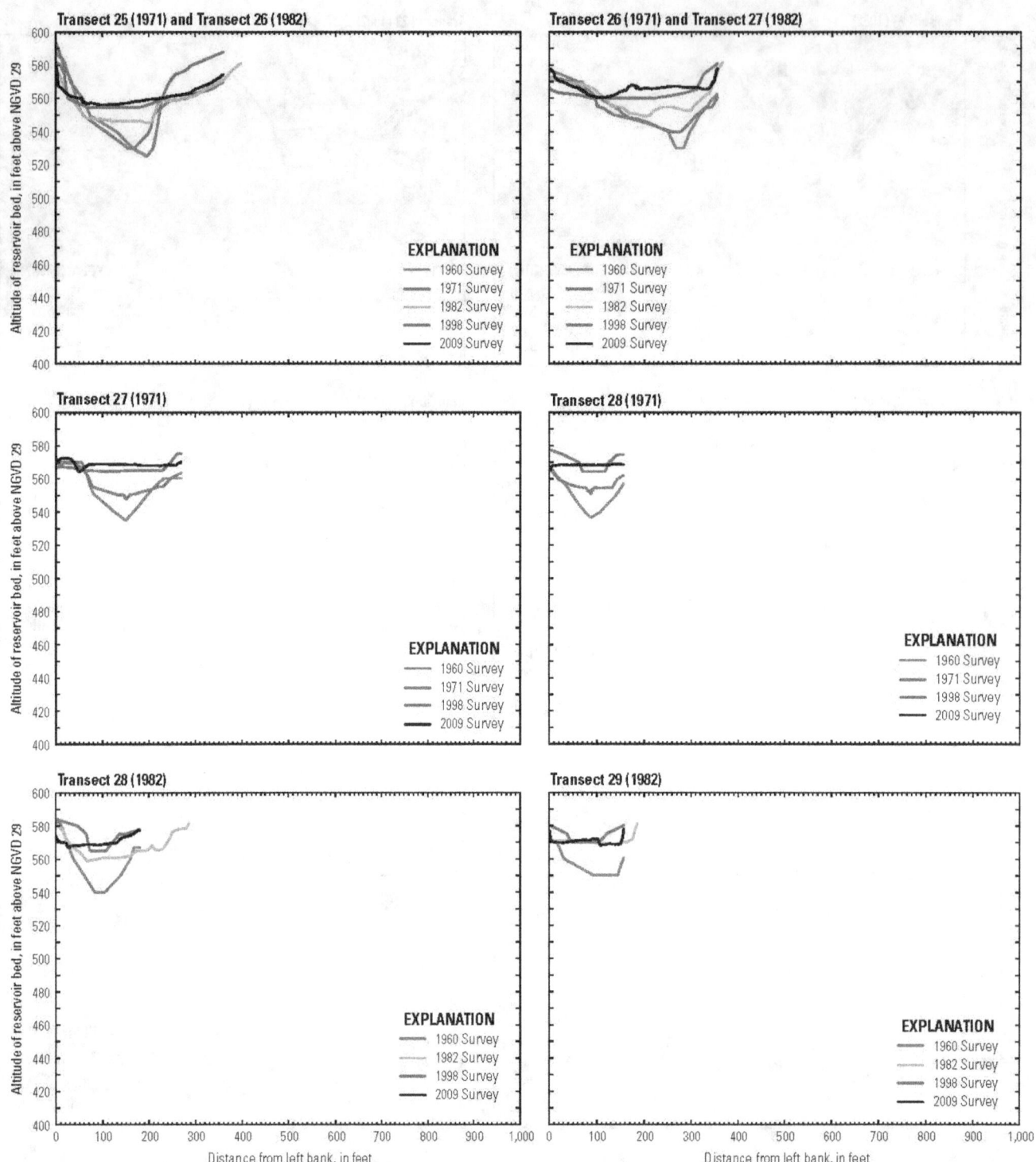

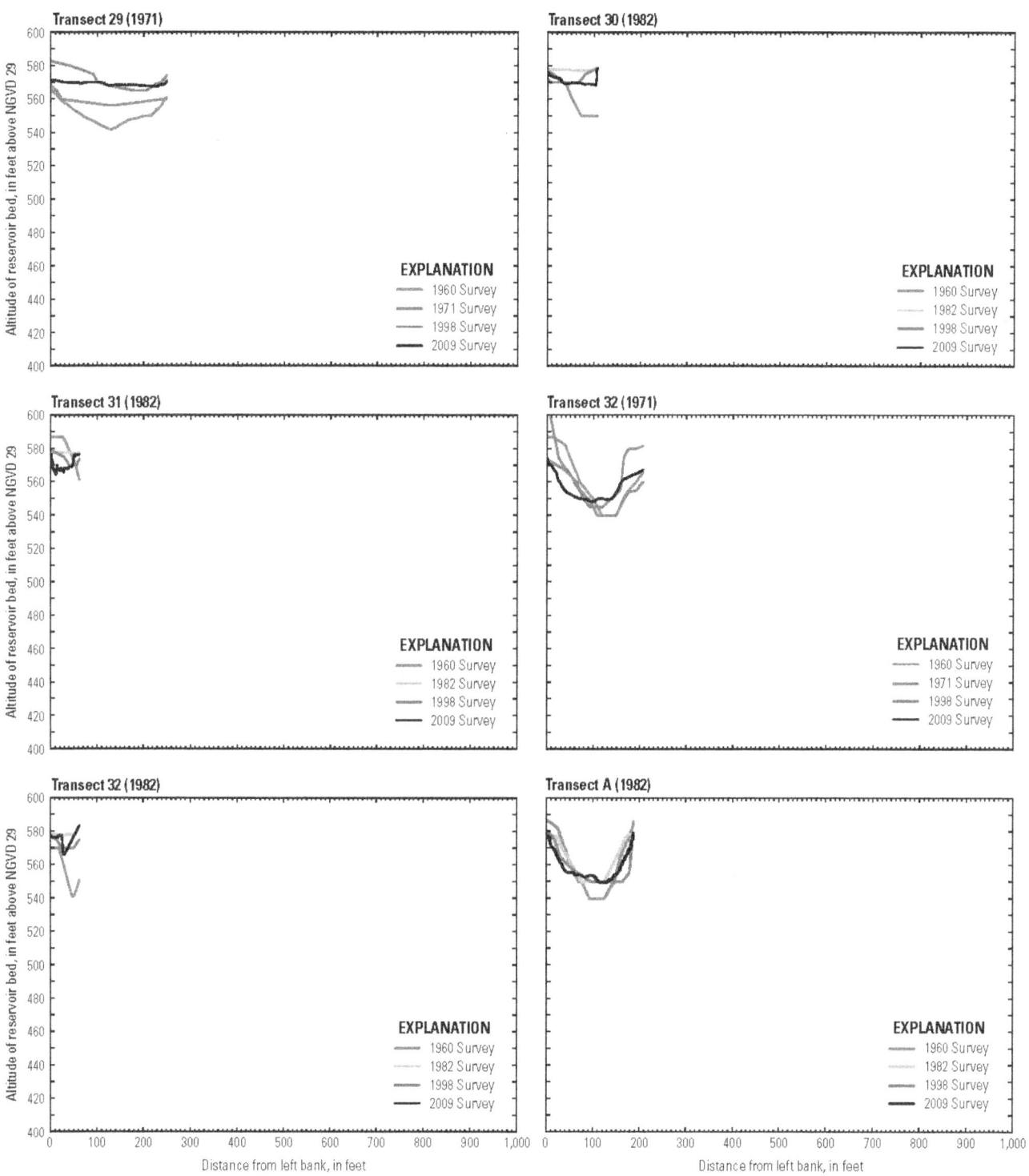

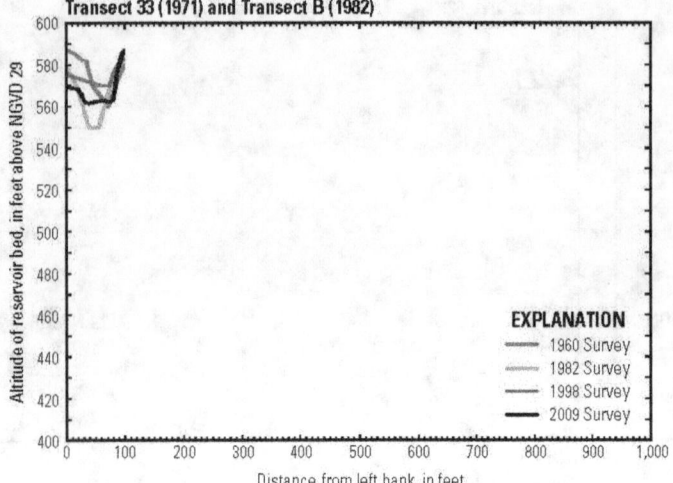

Transect 33 (1971) and Transect B (1982)